Biochemistry and Molecular Biology of Antimicrobial Drug Action

Biochemistry and Molecular Biology of Antimicrobial Drug Action

Sixth edition

T.J. Franklin and G.A. Snow

AstraZeneca
Alderley Park, Macclesfield
Cheshire
England

Springer

Library of Congress Cataloging-in-Publication Data

A C.I.P. record for this book is available from the Library of Congress.

ISBN 0-387-22554-4

Franklin and Snow: Biochemistry and Molecular Biology of Antimicrobial Drug Action

First Indian Reprint 2008

ISBN 978-81-8128-924-7

This edition is manufactured in India for sale only in India, Pakistan, Bangladesh, Nepal and Sri Lanka and any other country as authorized by the publisher.

This edition is published by Springer (India) Private Limited,
A part of Springer Science+Business Media, Registered Office: 906-907,
Akash Deep Building, Barakhamba Road, New Delhi – 110 001, India.

Printed in India by Saurabh Printers Private Ltd., Noida.

This edition is dedicated to the memory of my friend and colleague, Dr. Terry D. Hennessey, 1934–1999.

*The image on the front cover shows the crystal structure of MexA, a periplasmic protein component of a drug efflux pump in Gram-negative bacteria. Each molecule of MexA is represented by a different color and contains a long alpha-helical coiled-coil and two beta-sheet rich domains. The image was kindly provided by Dr Mark Higgins and his colleagues of the Medical Research Council Laboratory of Molecular Biology, Cambridge, UK and the Department of Pathology, Cambridge University. Full details of the structure are available in: Proceedings of the National Academy of Sciences, vol.**101**, pp. 9994–9999, (2004).*

Contents

Preface

The rapid advances made in the study of the synthesis, structure and function of biological macromolecules in the last fifteen years have enabled scientists concerned with antimicrobial agents to achieve a considerable measure of understanding of how these substances inhibit cell growth and division. The use of antimicrobial agents as highly specific inhibitors has in turn substantially assisted the investigation of complex biochemical processes. The literature in this field is so extensive, however, that we considered an attempt should be made to draw together in an introductory book the more significant studies of recent years. This book, which is in fact based on lecture courses given by us to undergraduates at Liverpool and Manchester Universities, is therefore intended as an introduction to the biochemistry of antimicrobial action for advanced students in many disciplines. We hope that it may also be useful to established scientists who are new to this area of research.

The book is concerned with a discussion of medically important antimicrobial compounds and also a number of agents that, although having no medical uses, have proved invaluable as research tools in biochemistry. Our aim has been to present the available information in a simple and readable way, emphasizing the established facts rather than more controversial material. Whenever possible, however, we have indicated the gaps in the present knowledge of the subject where further information is required. We have avoided the use of literature references in the text; instead we have included short lists of key articles and books for further reading at the end of each chapter.

We have drawn on the work of many scientists and we are especially pleased to express our thanks to those who have given us permission to reproduce their original diagrams and photographs. We are also grateful to the Pharmaceuticals Division of Imperial Chemicals Industries Ltd, for providing the necessary facilities for the preparation of this book.

Abbreviations used without definition for common biochemical substances are those recommended by the *Biochemical Journal* (1970).

T. J. FRANKLIN
G. A. SNOW

June 1970

Preface to the sixth edition

Since the publication of the previous edition, the problems posed by infectious diseases afflicting human beings and their domestic animals have continued to attract worldwide concern. The menace of AIDS remains unabated and is of epidemic proportions in parts of the developing world. The spread of multidrug-resistant bacteria, sporadic outbreaks of meningitis, bacterially mediated food poisoning and dangerous new viral infections regularly alarm the public. The mysterious emergence of bovine spongiform encephalitis, or mad cow disease, and its human equivalent, new variant Creuzfeldt-Jacob disease, presents a major challenge to medical and veterinary science. The potential of both wild-type and genetically modified infectious micro-organisms for bioterrorism is especially worrying. Fortunately, against this rather gloomy picture can be set some significant advances. Rapid nucleic acid and protein sequencing technology, sophisticated computer software to organize and analyze the huge amounts of emerging sequence data and spectacular advances in the X-ray crystallographic and nuclear magnetic resonsance spectroscopic investigation of macromolecular structures have all contributed to advances in our understanding of the mechanisms of antimicrobial action and drug resistance. These encouraging developments should facilitate the discovery of new drugs. Some valuable new antimicrobial drugs have emerged, including novel agents against influenza and further developments in the treatment of AIDS where combinations of anti-HIV drugs continue to bring hope to victims of this appalling disease.

Sadly, my former co-author, Alan Snow, died in 1995 and although I must therefore take sole responsibility for the contents of this new edition, I have been greatly helped by the incisive comments of the following scientists; Drs. Boudewijn de Jonge, John Rosamond, Thomas A. Keating, Peter Doig, Ann Eakin and Wright W. Nichols. Dr. Paul M. O'Neill of Liverpool University kindly provided me with an advance copy of a comprehensive review of artemisinin and related endoperoxides co-authored by himself and Dr. Gary H. Posner of The Johns Hopkins University. Over the years many helpful comments and criticisms from our readers have been invaluable in planning future editions. I hope they will continue to let me have their views. As in the previous editions, the discussions are mainly concerned with antimicrobial agents in medical, industrial and domestic use.

Finally, I would like to express my thanks to AstraZeneca for the provision of facilities which have made this new edition possible and especially to the Mereside Library staff for their help and advice on many occasions.

TREVOR J. FRANKLIN

Chapter one

The development of antimicrobial agents, past, present and future

1.1 The social and economic importance of antimicrobial agents

Few developments in the history of medicine have had such a profound effect upon human life and society as the development of the power to control infections by micro-organisms. In 1969 the Surgeon General of the United States stated that it was time 'to close the book on infectious diseases'. His optimism, which was then shared by many, seemed justified at the time. In the fight against infectious disease, several factors had combined to produce remarkable achievements. The first advances were mainly the result of improved sanitation and housing. These removed some of the worst foci of infectious disease and limited the spread of infection through vermin and insect parasites or by contaminated water and food. The earliest effective direct control of infectious diseases was achieved through vaccination and similar immunological methods which still play an important part in the control of infection today. The use of antimicrobial drugs for the control of infection was almost entirely a development of the twentieth century, and the most dramatic developments have taken place only since the 1930s. Surgery is no longer the desperate gamble with human life it had been in the early nineteenth century. By the late nineteenth century, the perils of childbirth had been greatly lessened with the control of puerperal fever. The death of children and young adults from meningitis, tuberculosis and septicemia, once commonplace, was, by the late 1960s, unusual in the developed world.

Unfortunately, since the heady optimism of the 1960s we have learned to our cost that microbial pathogens still have the capacity to spring unpleasant surprises on the world. The problem of acquired bacterial resistance to drugs, recognized since the very beginning of antimicrobial chemotherapy, has become ever more menacing. Infections caused by the tubercle bacillus and *Staphylococcus aureus*, which were once readily cured by drug therapy, are now increasingly difficult or even impossible to treat because of widespread bacterial resistance to the available drugs. Nor is resistance confined to these organisms; many other species of bacteria as well as fungal pathogens, viruses and protozoa, have also become drug-resistant. The ability of micro-organisms to kill or disable the more vulnerable members of society, especially the very young and old and patients with weakened immune defences, is reported in the media almost daily. Alarming reports of lethal enteric infections, meningitis and 'flesh-eating' bacteria have become depressingly familiar. If this were not enough, the spectre of the virus (HIV) infection which leads to AIDS (acquired immune deficiency syndrome) threatens human populations around the world, in nations both rich and poor.

Drug therapy for AIDS can, initially at least, be very effective, but for the occasional outbreaks of terrifying viral infections such as Ebola and Lassa fever, there are no treatments,. Perhaps more worrying than these sporadic African hemorrahgic fevers, however, is the perceived risk of epidemic or pandemic infections caused by the recently discovered severe acute respiratory syndrome (SARS) virus, or novel recombinations of highly virulent influenza viruses with the potential for causing severe illness and death on a catastrophic scale. In recent years the mosquito-borne West Nile virus, which in some cases can cause a potentially fatal encephalitis, has been the subject of increasing concern in North America. Throughout much of the tropical and subtropical world, malaria continues to exact a dreadful toll on the health and lives of inhabitants. Although mass movements of populations and the failure to control the anophelene mosquito insect vector are major factors in the prevalence of malaria, the increasing resistance of the malarial protozoal parasite to drug treatment is of the greatest concern.

Thirty years ago serious infections caused by fungi were relatively rare. More common infections like thrush and ringworm were more of an unpleasant nuisance than a serious threat to health. Today, however, many patients with impaired immunity caused by HIV infection, cytotoxic chemotherapy for malignant disease, or the immunosuppressive treatment associated with organ graft surgery, are at risk from dangerous fungal pathogens such as *Pneumocystis carinii* and *Cryptococcus neoformans*. Less virulent organisms like *Candida albicans* can also be devastating in immunocompromised patients. Inevitably, the increasing use of antifungal drugs to control these infections results in the emergence of drug-resistant pathogens.

Fortunately, despite the threats posed by drug-resistant bacteria, viruses, protozoal parasites and fungal pathogens, the current scene is not one of unrelieved gloom. Most bacterial and fungal infections can still be treated successfully with the remarkable array of drugs available to the medical (and veterinary) professions. Work continues to develop drugs effective against resistant pathogens, and there has been major progress against viruses causing AIDS, influenza and herpes infections. Vaccines are remarkably successful in preventing some bacterial and viral infections. Indeed, outstanding amongst the medical achievements of the twentieth century was the eradication of smallpox and the dramatic reduction in the incidence of poliomyelitis through mass vaccination programmes. A further incentive to the discovery and development of novel antimicrobial drugs and vaccines is the threat of bioterorrism, which could exploit conventional lethal pathogens such as anthrax and smallpox or even micro-organisms genetically manipulated to extraordinary levels of virulence and drug resistance.

Finally, mention must be made of the recent and unexpected emergence of infectious prions which are associated with such devastating neurological pathologies as Creuzfeldt-Jacob disease (CJD) and new variant CJD, or mad cow disease in humans. It is now almost universally accepted that the heat and chemically resistant prion particles are proteinaceous and contain no detectable nucleic acid-encoded information. Infection is transmitted by various routes, including oral ingestion, for example, in contaminated food, by injection, or during surgery and possibly by blood transfusion. At present there is no effective drug treatment to arrest or delay the relentless progression of the infection, which involves the conversion of a normal neuronal protein of unknown function to an insoluble and newly infectious form through interaction with the invading, closely related prion protein. The disturbing possibility of a slowly developing epidemic of new variant CJD is spurring efforts to find drugs or vaccines to control prion infections.

1.2 An outline of the historical development of antimicrobial agents

1.2.1 Early remedies

Among many traditional and folk remedies, three sources of antimicrobial compounds have survived to the present day. These are cinchona bark and *Artemisia annua* (Chinese quing hao su) for the treatment of malaria and ipecacuanha root for amebic dysentery. Cinchona bark was used by the Indians of Peru for treating malaria and was introduced into European medicine by the Spanish in the early seventeenth century. The active principle, quinine, was isolated in 1820. Quinine remained the only treatment for malaria until well into the twentieth century and still has a place in chemotherapy. The isolation of artemisinin, the active compound in *Artemisia annua*, by Chinese scientists is much more recent and only in recent years has its therapeutic potential against malaria been fully appreciated. Ipecacuanha root was known in Brazil and probably in Asia for its curative action in diarrheas and dysentery. Emetine was isolated as the active constituent in 1817 and was shown in 1891 to have a specific action against amebic dysentery. In combination with other drugs, it is still used for treating this disease. These early remedies were used without any understanding of the nature of the diseases. Malaria, for example, was thought to be caused by 'bad air' (mal'aria) arising from marshy places; the significance of the blood-borne parasite was not recognized until 1880, and only in 1897 was the anophelene mosquito proved to be the specific insect vector when the developing parasite was observed in the intestine of the mosquito.

1.2.2 Antiseptics and disinfectants

The use of disinfectants and antiseptics also preceded an understanding of their action, and seems to have arisen from the observation that certain substances stopped the putrefaction of meat or the rotting of wood. The term 'antiseptic' itself was apparently first used by Pringle in 1750 to describe substances that prevent putrefaction. The idea was eventually applied to the treatment of suppurating wounds. Mercuric chloride was used by Arabian physicians in the Middle Ages to prevent sepsis in open wounds. However, it was not until the nineteenth century that antiseptics came into general use in medicine. Chlorinated soda, essentially hypochlorite, was introduced in 1825 by Labarraque for the treatment of infected wounds, and tincture of iodine was first used in 1839. One of the earliest examples of disinfection used in preventing the spread of infectious disease was recorded by Oliver Wendell Holmes in 1835. He regularly washed his hands in a solution of chloride of lime when dealing with cases of puerperal fever and thereby greatly reduced the incidence of fresh infections, as did Ignaz Semmelweiss in Vienna a few years later. These pioneer attempts at antisepsis were not generally accepted until Pasteur's publication in 1863 identifying the microbial origin of putrefaction. This led to an understanding of the origin of infection and suggested the rationale for its prevention. As so often in the history of medicine, a change in practice depended upon the personality and persistence of one man. In antiseptics this man was Lister. He chose phenol, the antiseptic which had been introduced by Lemaire in 1860, and applied it vigorously in surgery. A 2.5% solution was used for dressing wounds and twice this concentration for sterilizing instruments. Later he used a spray of phenol solution to produce an essentially sterile environment for carrying out surgical operations. The previous state of surgery had been deplorable; wounds usually became infected and the mortality was appalling. The effect of Lister's measures was revolutionary, and the antiseptic technique opened the way to great surgical advances. Even at this time, about 1870, the use of antiseptics was still empirical. An understanding of their function began with the work of Koch, who from 1881 onwards introduced the techniques on which modern bacteriology has been built. He perfected methods of obtaining pure cultures of bacteria and growing them on solid media, and he demonstrated practical methods of sterile working. Once it became possible to handle bacteria in a controlled environment, the action of disinfectants and antiseptics could be studied. The pioneer work on the scientific approach to this subject was published by Kronig and Paul in 1897.

Since that time, the history of antiseptics has been one of steady but unspectacular improvement. Many of the traditional antiseptics have continued in use in refined forms. The phenols have been modified and made more acceptable for general use. Acriflavine, introduced in 1913, was the first of a number of basic antiseptics. It had many years of use but was displaced by colourless cationic antiseptics and the non-ionic triclosan (acriflavine is bright orange). In surgery the antiseptic era gave way to the aseptic era in which the emphasis is on the avoidance of bacterial contamination rather than on killing bacteria already present. All the same, infection of surgical wounds remains a constant risk and antiseptics are still used as an extra precaution or second line of defence. Surgical staff also 'scrub up' with mild antiseptic solutions before entering the operating theatre. Disinfectants play an important part in the hygiene of milking sheds, broiler houses and other places where strict asepsis is impracticable.

1.2.3 The beginnings of chemotherapy

The publications of Pasteur and Koch firmly established that micro-organisms are the cause of infectious disease, though for some diseases the causative organisms still remained to be discovered. It was also known that bacteria are killed by various antiseptics and disinfectants. Not surprisingly, attempts were made to kill micro-organisms within the body and so end the infection. Koch himself carried out some experiments with this aim. He had shown that mercuric chloride is one of the few disinfectants that kill the particularly tough spores of the anthrax bacillus. Koch therefore tried to cure animals of anthrax infection by injecting them with mercuric chloride. Unfortunately the animals died of mercury poisoning and their organs still contained infectious anthrax bacilli. A slightly more successful attempt to cure an infection with a toxic agent was made by Lindgard in 1893. He treated horses suffering from surra, a disease now known to be caused by trypanosomes, with arsenious oxide. There was some improvement of the disease, but the compound was too toxic to be generally useful.

Chemotherapy, however, really began with Paul Ehrlich. During the ten years from 1902 onwards Ehrlich's work foreshadowed many of the concepts which have governed subsequent work on antimicrobial agents. His first ideas arose from studies with 'vital stains', dyestuffs that were taken up selectively by living tissue. One such dye was methylene blue, which in the animal body is concentrated in nervous tissue. Ehrlich showed that the same dye is readily taken up by malaria parasites in the blood so that they become deeply stained. Consequently methylene blue was tried against human malaria and showed some effect, though not sufficient to make it a useful treatment. Nevertheless, this minor success started a line of thought that was to prove of the greatest significance. Ehrlich believed that antimicrobial agents must be essentially toxic compounds and that they must bind to the micro-organism in order to exert their action. The problem was to discover compounds having a selective action against the microbial cell rather than the cells of the host animal. Starting from methylene blue, Ehrlich began to search for other dyestuffs that would affect protozoal diseases. In 1904, after testing hundreds of available dyes, he eventually found one that was effective against trypanosomiasis in horses. This compound, called trypan red, was a significant landmark in the treatment of microbial infections since it was the first man-made compound that produced a curative effect.

However, it was not in the field of dyestuffs that Ehrlich achieved his greatest success. Following the early work on the treatment of trypanosomiasis with arsenious oxide, Koch tested the organic arsenical, atoxyl (Figure 1.1). This compound produced the first cures of sleeping sickness, a human trypanosomal disease. Unfortunately, however, the compound produced serious side effects, with some patients developing optic atrophy. The curative effect of this compound stimulated Ehrlich to make other related arsenicals. He tested these on mice infected experimentally with trypanosomiasis and showed that curative action did not parallel toxicity to the mice. This suggested that if enough compounds were made, some would have sufficiently low toxicity to be safe as chemotherapeutic agents. Ehrlich continued his search for compounds active against various micro-organisms and showed that arsenicals were active against the causative organism of syphilis. He began a massive search for an organoarsenical compound that could be used in the treatment of this disease and eventually in 1910 discovered the famous drug salvarsan (Figure 1.1). This

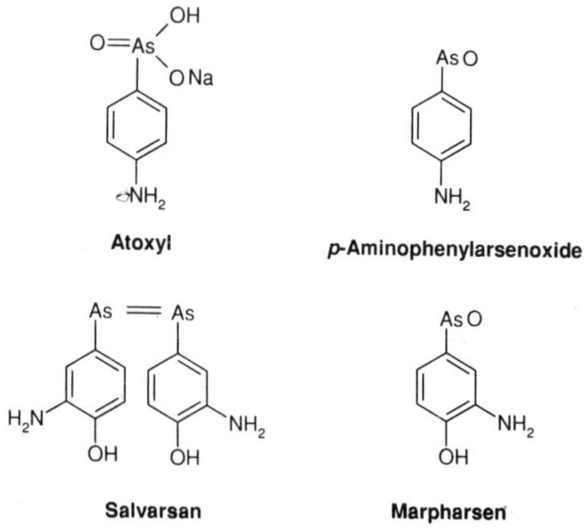

FIGURE 1.1 Arsenical compounds used in the early treatment of trypanosomiasis or syphilis.

drug and its derivative, Neosalvarsan, became the standard treatment for syphilis. Coupled with bismuth therapy, they remained in use until supplanted by penicillin in 1945. This was the most spectacular practical achievement of Ehrlich's career, but scientifically he is remembered at least as much for his wealth of ideas that have inspired workers in the field of chemotherapy down to the present day. These ideas are so important that they deserve separate consideration.

1.2.4 The debt of chemotherapy to Ehrlich

The very term 'chemotherapy' was invented by Ehrlich and expressed his belief that infectious disease could be controlled by treatment with synthetic chemicals. Successes since his day have entirely justified his faith in this possibility. He postulated that cells possess chemical receptors which are concerned with the uptake of nutrients. Drugs that affect the cell must bind to one or other of these receptors. The toxicity of a drug is determined partly by its distribution in the body. However, in the treatment of an infection, the binding to the parasite relative to the host cell determines the effectiveness of the compound. Thus Ehrlich recognized the importance of quantitative measurement of the relationship between the dose of a compound required to produce a therapeutic effect and the dose that causes toxic reactions. Such measurements are still of prime importance in chemotherapy today.

Ehrlich pioneered methods that have since become a mainstay of the search for new drugs. One aspect of his approach was the use of screening. This is the application of a relatively simple test to large numbers of compounds in order to obtain evidence of biological activity in types of chemical structure not previously examined. Modern drug-discovery laboratories usually employ sequences or cascades of screening tests, often beginning with a purified enzyme from the target organism, followed by test cultures of the pathogen and sophisticated evaluation in experimental animals. Vast numbers of compounds may be screened in the primary *in vitro* test, with the succeeding screens used to progressively filter out insufficiently active or potentially toxic compounds until a very limited set of compounds is considered to be sufficiently effective to warrant evaluation in a model infection in experimental animals.

The second of Ehrlich's methods was the synthesis of chemical variants of a compound exhibiting an interesting but not optimal level of activity. The new compounds were examined for increased activity or for improvements in some other property, such as reduced toxicity. Any improvement found was used as a guide to further synthesis, eventually arriving by a series of steps at the best possible compound. These methods are now so well accepted that their novelty in Ehrlich's day can easily be forgotten. They depend on the thesis that a useful drug possesses an ideal combination of structural features which cannot be predicted at the outset. A compound approximating this ideal will show some degree of activity and can therefore act as a 'lead' towards the best attainable structure.

According to Ehrlich, a chemotherapeutic substance has two functional features, the 'haptophore' or binding group which enables the compound to attach itself to the cell receptors, and the 'toxophore' or toxic group which brings about an adverse effect on the cell. This idea has had a continuing influence in subsequent years. In cancer chemotherapy it has frequently been used in attempts to bring about the specific concentration of toxic agents or antimetabolites in tumour cells. In antimicrobial research it has helped to explain some features of the biochemical action of antimicrobial compounds.

Ehrlich also recognized that compounds acting on microbial infection need not necessarily kill the invading organism. It was, he suggested, sufficient to prevent substantial multiplication of the infectious agent, since the normal body defences, antibodies and phagocytes, would cope with foreign organisms provided that their numbers were not overwhelming. His views on this topic were based in part on his observation that isolated syphilis-causing spirochetes treated with low concentrations of salvarsan remained motile and were therefore apparently still alive. Nevertheless they were unable to produce an infection when they were injected into an animal body. It is a striking fact that several of today's important antibacterial and antifungal drugs are 'biostatic' rather than 'biocidal' in action.

Another feature of Ehrlich's work was his recognition of the possibility that drugs may be activated by metabolism in the body This suggestion was prompted by the observation that the compound atoxyl was active against trypanosomal infections but was inactive against isolated trypanosomes. His explanation was that atoxyl was reduced in the body to the much more toxic *p*-aminophenylarsenoxide (Figure 1.1). Later work showed that atoxyl and other related arsenic acids are not in fact readily reduced to arsenoxides in the body, but local reduction by the parasite remains a possibility. Ehrlich, surprisingly, did not recognize that his own compound salvarsan would undergo metabolic cleavage. In animals it gives rise to the arsenoxide as the first of a series of metabolites. This compound eventually was introduced into medicine in 1932 under the name Mapharsen (Figure 1.1); its toxcity is rather high, but it has sufficient selectivity to give it useful chemotherapeutic properties. Other examples of activation through metabolism have been discovered in more recent times; for example, the conversion of the antimalarial 'prodrug' proguanil to the active cycloguanil in the liver and the metabolism of antiviral nucleosides to the inhibitory triphosphate derivatives. Of course, it later emerged that metabolism in the body or in the infecting micro-organism could also result in the inactivation of drugs.

Ehrlich also drew attention to the problem of resistance of micro-organisms towards chemotherapeutic compounds. He noticed it in the treatment of trypanosomes with parafuchsin and later with trypan red and atoxyl. He found that resistance extended to other compounds chemically related to the original three, but there was no cross-resistance among the groups. In Ehrlich's view this was evidence that each of these compounds was affecting a separate receptor. Independent resistance to different drugs later became a commonplace in antimicrobial therapy. Ehrlich's view of the nature of resistance is also interesting. He found that trypanosomes resistant to trypan red absorbed less of the compound than sensitive strains, and he postulated that the receptors in resistant cells had a diminished affinity for the dye. This mechanism corresponds to one of the currently accepted types of resistance in micro-organisms (Chapter 9) in which mutations affecting the target protein reduce or eliminate drug binding to the target.

Several useful antimicrobial drugs appeared in later years as an extension of Ehrlich's work. The most notable (Figure 1.2) are suramin, developed from trypan red, and mepacrine (also known as quinacrine or atebrin) developed indirectly from methylene blue (Figure 1.2). Suramin, introduced in 1920, is a colorless compound with useful activity against human trypanosomiasis. Its particular value lay in its relative safety compared with other antimicrobial drugs of the period. It was the first useful antimicrobial drug without a toxic metal atom, and the ratio of the dose required to produce toxic symptoms to that needed for a curative effect is vastly higher than with any of the arsenicals. Suramin is remarkably persistent, a single dose giving protection for more than a month. Mepacrine, first marketed in 1933, was an antimalarial agent of immense value in the Second World War. It was supplanted by other compounds partly because it caused a yellow discoloration of the skin. Besides these obvious descendants from Ehrlich's work, the whole field of drug therapy is permeated by his ideas, and many other important compounds can be traced directly or indirectly to the influence of his thought.

1.2.5 The treatment of bacterial infections by synthetic compounds

In spite of the successes achieved in the treatment of protozoal diseases and the spirochetal disease syphilis, the therapy of bacterial infections remained for many years an elusive and apparently unattainable goal. Ehrlich himself, in collaboration with Bechtold, made a series of phenols which showed much higher antibacterial potency than the simple phenols originally used as disinfectants. These compounds, however, had no effect on bacterial infections in animals. Other attempts were equally unsuccessful and no practical progress was made until 1935, when Domagk reported the activity of prontosil rubrum (Figure 4.1) against infections in animals. The discovery occurred in the course of a widespread research programme on the therapeutic use of dyestuffs, apparently inspired by Ehrlich's ideas. Trefouel showed that prontosil rubrum is broken down in the body, giving sulfanilamide (Figure 4.1), which was in fact the effective antibacterial agent. The sulfonamides might have been developed and used more widely if penicillin and other antibiotics had not followed on so speedily. In fact, relatively

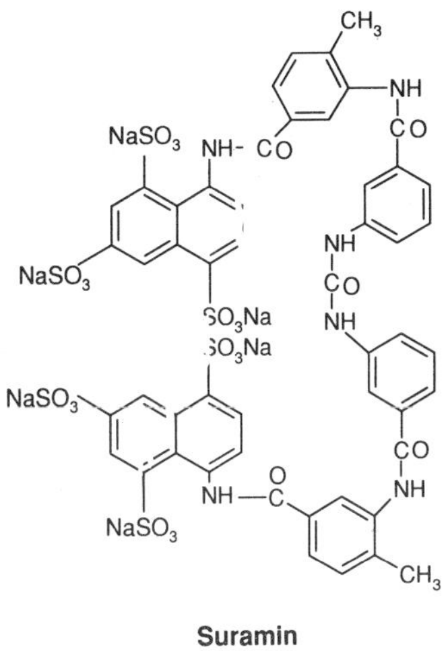

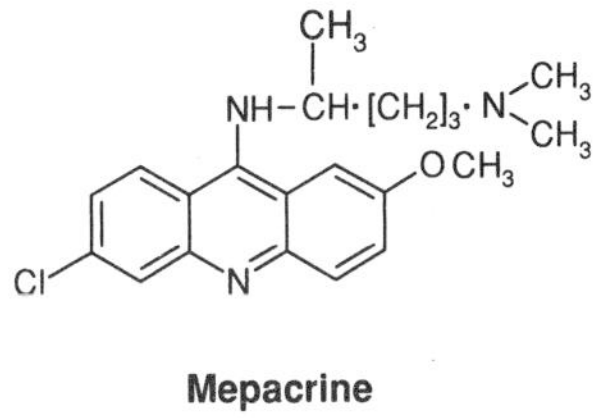

FIGURE 1.2 Early synthetic compounds used for treating protozoal infections: suramin for trypanosomiasis (African sleeping sickness) and mepacrine for malaria.

few wholly synthetic compounds have achieved success against the common bacterial infections mainly because effective synthetic compounds have been difficult to find. After some 60 years of effort, the synthetic antibacterial compounds in current use include, in addition to a few sulfonamides, several drugs for the treatment of tuberculosis, the quinolones, trimethoprim, certain imidazoles such as metronidazole, nitrofurans such as nitrofurantoin, and most recently the oxazolidinone, linezolid. There are also numerous semisynthetic derivatives of naturally occurring antibacterial antibiotics, including β-lactams, aminoglycosides, macrolides, streptogramins and glycopeptides.

For several years after treatment was available for streptococcal and staphylococcal infections, the mycobacterial infections that cause tuberculosis and leprosy remained untreatable by chemotherapy. The first success came with the antibiotic streptomycin, which remains an optional part of the standard treatment for tuberculosis. Later, several chemically unrelated synthetic agents were also found to be effective against this disease. The best of these are isonicotinyl hydrazide (isoniazid), ethambutol and pyrazinamide. Another antibiotic, rifampicin (rifampin), is usually included in the current combination therapy for tuberculosis. The synthetic compound, 4,4′-diaminodiphenylsulfone, is regularly used in the treatment of leprosy.

1.2.6 The antibiotic revolution

Ever since bacteria have been cultivated on solid media, contaminant organisms have occasionally appeared on the plate. Sometimes this foreign colony would be surrounded by an area in which bacterial growth was suppressed. Usually this was regarded as a mere technical nuisance, but in 1928, observing such an effect with the mold *Penicillium notatum* on a plate seeded with staphylococci, Alexander Fleming was struck by its potential importance. He showed that the mold produced a freely diffusible substance highly active against Gram-positive bacteria and apparently of low toxicity to animals. He named it penicillin. It was, however, unstable and early attempts to extract it failed, so Fleming's observation lay neglected until 1939. By then the success of the sulfonamides had stimulated a renewed interest in the chemotherapy of bacterial infections. The search for other antibacterial agents now seemed a promising and exciting project, and Howard Florey and Ernst Chain selected Fleming's penicillin for re-examination. They succeeded in isolating an impure but highly active solid preparation and published their results in 1940. Evidence of its great clinical usefulness in patients followed in 1941. Because of the extraordinary antibacterial potency of penicillin and its minimal toxicity to patients, it was apparent that a compound of revolutionary importance in medicine had been discovered. Making it generally available for medical use, however, presented formidable problems both in research and in large-scale production, especially under conditions of wartime stringency. Eventually perhaps the biggest chemical and biological joint research programme ever mounted was undertaken, involving 39 laboratories in Britain and the United States. It was an untidy operation with much duplication and overlapping of work, but it culminated in the isolation of pure penicillin, the determination of its structure, and the establishment of the method for its production on a large scale. The obstacles overcome in this research were enormous. They arose mainly from the very low concentrations of penicillin in the original mold cultures and from the marked chemical instability of the product. In the course of this work the concentration of penicillin in mold culture fluids was increased 1000-fold by the isolation of improved variants of *Penicillum notatum* using selection and mutation methods and by improved conditions of culture. This tremendous improvement in yield was decisive in making large-scale production practicable and ultimately cheap.

The success of penicillin quickly diverted a great deal of scientific effort towards the search for other antibiotics. The most prominent name in this development was that of Selman Waksman, who began an intensive search for antibiotics in micro-organisms isolated from soil samples obtained in many parts of the world. Waksman's first success was streptomycin, and other antibiotics soon followed. Waksman's technique of screening soil organisms for antibiotics was immediately copied in many other laboratories. Organisms of all kinds were examined and hundreds of thousands of cultures were tested. Further successes came quickly. Out of all this research, several thousand named antibiotics have been listed. Most of them, however, have adverse properties that prevent their development as drugs. Perhaps 50 have had some sort of clinical use and only a few of these are regularly employed in the therapy of infectious disease. However, among this select group and their semisynthetic variants are compounds of such excellent qualities that treatment is now available for most of the bacterial infections known to occur in humans, although as we have seen, resistance increasingly threatens the efficacy of drug therapy.

Following the wave of discovery of novel classes of antibiotics in the 1940s and 1950s, research focused largely on taking antibiotics of proven worth and subjecting them to chemical modification in order to extend their antibacterial spectrum, to combat resistance and to improve their acceptability to patients. Recently, however, the pressure of increasing drug resistance has renewed efforts to discover novel chemical classes of both naturally occurring and synthetic antibacterial compounds.

1.2.7 Antifungal and antiviral drugs

The diversity of fungal pathogens which attack man and his domesticated animals is considerably smaller than that of bacteria. Nevertheless, fungi cause infections ranging from the trivial and inconvenient to those resulting in major illness and death. Fungal infections have assumed greater importance in recent years because of the increased number of medical conditions in which host immunity is compromised. Fungi as eukaryotes have much more biochemistry in common with mammalian cells than bacteria do and therefore pose a serious challenge to chemotherapy. Specificity of action is more difficult to achieve. Few antibiotics are useful against fungal infections, and attention has concentrated on devising synthetic agents. Some advantage has been taken of the progress in producing compounds for the treatment of fungal infections of plants to develop from them reasonably safe and effective drugs for human fungal infections.

Enormous strides have been made in the control of viral infections through the use of vaccines. Smallpox has been eradicated throughout the world. In the developed countries at least, the seasonal epidemics of poliomyelitis that were the cause of so much fear and suffering 50 years ago have disappeared. But despite these and other vaccine-based successes against viral infections, not all such infections can be so effectively controlled by mass vaccination programmes. The bewildering diversity of common cold viruses, the ever-shifting antigenic profiles of influenza viruses and the insidious nature of the virus that leads to AIDS are just three examples of diseases that may not yield readily to the vaccine approach. Attention is therefore focused on finding drugs that specifically arrest or prevent viral infection, a formidable challenge since viruses partially parasitize the biochemistry of the host cells. Nevertheless, considerable success has been achieved in devising effective drugs against several viruses, including HIV, herpes and cytomegalo-viruses and even against influenza viruses. Recombinant forms of the naturally occurring antiviral protein interferon-α (IFN-α), have a useful role in combating the viruses which cause the liver infections hepatitis B and C.

1.2.8 Antiprotozoal drugs

After the Second World War, several valuable new drugs were introduced in the fight against malaria, including chloroquine, proguanil and pyrimethamine. For several years these drugs were extremely effective for both the prevention and treatment of malaria. However, by the time of the outbreak of the war in Vietnam in the 1960s it had become clear that, like bacteria, the malarial parasites were adept at finding ways to resist drug therapy. The US government then launched a massive screening project to discover new antimalarial agents. Two compounds, mefloquine and halofantrine, resulted from this effort and are still in use today. Nevertheless, the development of resistance to these drugs seems inevitable and the search for new antimalarial drugs continues. Several compounds currently offer real promise: the naturally occurring compound, artemisinin and its semisynthetic derivatives, and the synthetic compound, atovaquone.

The treatment of other serious protozoal infections, such as the African and South American forms of trypanosomiasis remains relatively primitive. The arsenical melarsoprol is still used for African trypanosomiasis (sleeping sickness), although the less toxic difluorodimethyl ornithine is increasingly seen as the drug of choice. South American trypanosomiasis, or Chagas' disease, is still very difficult to treat successfully; control of the insect vector, the so-called kissing bug which infests poor-quality housing, is very effective. The only useful drugs against leishmaniasis are such venerable compounds as sodium antimony gluconate and pentamidine, neither of which is ideal. Unfortunately, the parasitic diseases of the developing world do not present the major pharmaceutical companies with attractive commercial opportunities, and research into the treatment of these diseases is relatively neglected.

1.3 Reasons for studying the biochemistry and molecular biology of antimicrobial compounds

Following this brief survey of the discovery of the present wide range of antimicrobial compounds, we may now turn to the main theme of the book. We shall be concerned with the biochemical mechanisms that underlie the action of compounds used in the battle against pathogenic micro-organisms. Where there is sufficient information, this will also include general descriptions of the interactions between drugs and their primary molecular targets. Increasingly, the detailed understanding of drug action at the molecular level is now used to generate ideas for the design of entirely novel antimicrobial agents. Antimicrobial agents, particularly the antibiotics, often have a highly selective action on biochemical processes. They may block a single reaction within a complex sequence of events. The use of such agents has often revealed details of biochemical processes that would otherwise have been difficult to disentangle. Attempts to understand the biochemistry of antimicrobial action were initially slow and painful, with many false starts and setbacks. Progress began to accelerate in the early 1960s and accompanied the dramatic advances being made at that time in the biochemistry and molecular biology of bacteria. In the past few years, truly remarkable developments in the genetic manipulation of bacteria, the plentiful production of hitherto inaccessible proteins and the elucidation of macomolecular structures by X-ray crystallography and nuclear magnetic resonance spectroscopy have taken our understanding of antibacterial drug action to unprecedented levels of detail.

Knowledge of the mechanism of action of antiprotozoal drugs, some of which were discovered long before the antibacterial drugs, lagged well behind for many years. This was due mainly to the difficulty in isolating and working with protozoa outside the animal body, but interest had also been concentrated on bacteria because of their special importance in infectious disease and their widespread use in biochemical and genetic research. However, advances in the molecular genetics of the major parasitic protozoans should now facilitate the development of our understanding of drug action in these species. Rapid progress is also being made in working out the biochemical and molecular basis of the action of antifungal and antiviral drugs.

1.4 Uncovering the molecular basis of antimicrobial action

Several steps in discovering the molecular basis of antimicrobial action can be distinguished and will be discussed separately.

1.4.1 Nature of the biochemical systems affected

As long as antimicrobial compounds have been known, scientists have attempted to explain their action in biochemical terms. Ehrlich made a tentative beginning in this direction when he suggested that the arsenicals might act by combining with thiol groups on the protozoal cells. He was, however, severely limited by the elementary state of biochemistry at that period. By the time the sulfonamides were discovered, the biochemistry of small molecules was more advanced and a reasonable explanation of the biochemical basis of sulfonamide action was soon available. However, many of the antibiotics which followed presented very different problems. Attempts to apply biochemical methods to the study of their action led to highly conflicting answers. At one stage a count showed that 14 different biochemical systems had been suggested as the site of action of streptomycin against bacteria. Much of this confusion arose from a failure to distinguish between primary and secondary effects. The biochemical processes of bacterial cells are closely interlinked. Thus disturbance of any one important system is likely to have effects on many of the others. Methods had to be developed that would distinguish between the primary biochemical effect of an antimicrobial agent and other changes in metabolism that followed as a consequence. Once these were established, more accurate assessments could be made of the real site of action of various antimicrobial compounds. The limiting factor then became the extent of biochemical information about the nature of the target site. From about 1960 onwards there have been continuing and remarkable advances in our understanding of the structure, function and synthesis of macromolecules. Most of the important antibiotics were found to act by interfering with the biosynthesis or function of macromolecules, and the development of new techniques provided the means of defining their site of action in ever-increasing detail.

1.4.2 Methods used to study the mode of action of antimicrobial compounds

Many of the early antimicrobial drugs were discovered by the simple method of screening for antimicrobial activity in collections of synthetic compounds and the media in which micro-organisms suspected of antibiotic production had been cultured. This approach provided little or no information as to the likely mechanism of antimicrobial action. However, experience over the past five decades has developed systematic procedures for working out the primary site of action for most of these empirically discovered compounds. Once the primary site of action is established, the overall effects of a drug on the metabolism of microbial cells can often be explained and the precise details of the interaction between the drug and its molecular target finally revealed. Many of the techniques are discussed in later chapters, but it may be helpful to set them out in a logical sequence.

1. The chemical structure of the drug is studied carefully to determine whether a structural analogy exists with part, or the whole, of a biologically important molecule, for example, a metabolic intermediate or essential cofactor, or nutrient. An analogy may be immediately obvious, but sometimes it becomes apparent only through imaginative molecular model building or by hindsight after the target site of the compound has been revealed by other means. This approach revealed the site of action of the sulfonamide antibacterial drugs. Nevertheless, analogies of structure

can sometimes be misleading and should only be used as a preliminary indication.

2. The next step is to examine the effects of the compound on the growth kinetics and morphology of suitable target cells. A cytocidal effect shown by a reduction in viable count may indicate damage to the cell membrane. This can be confirmed by observation of leakage of potassium ions, nucleotides or amino acids from the cells. Severe damage leads to cell lysis. Examination of bacterial and fungal cells by electron microscopy may show morphological changes which indicate interference with the synthesis of one of the components of the cell wall. Many antibiotics have only a cytostatic action and do not cause any detectable morphological changes.
3. Usually attempts are made to reverse the action of an inhibitor by adding various supplements to the medium. Cellular nutrients, including oxidizable carbon sources, fatty acids, amino acids, the nucleic acid precursors purines and pyrimidines, and vitamins are tested in turn. If reversal is achieved, this may point to the reaction or reaction sequence which is blocked by the inhibitor. Valuable confirmatory evidence can often be obtained by the use of genetically engineered auxotrophic organisms which require a compound known to be the next intermediate in a biosynthetic sequence beyond the reaction blocked by the antimicrobial agent. Auxotrophs of this type should be resistant to the action of the inhibitor. Inhibition in a biosynthetic sequence may also be revealed by accumulation of the metabolite immediately before the blocked reaction. Unfortunately, the actions of many antimicrobial agents are not reversed by exogenous compounds. This especially applies to compounds which interfere with the polymerization stages in nucleic acid and protein biosynthesis, where reversal is impossible.
4. The ability of an inhibitor to interfere with the supply and consumption of ATP is usually examined since any disturbance of energy metabolism has profound effects on the biological activity of the cell. The inhibitor is tested against the respiratory and glycolytic activities of the micro-organism, and the ATP content of the cells is measured. Compounds which damage biological membranes are likely to collapse the proton gradient across the cytoplasmic membrane of bacteria and thereby block the biosynthesis of ATP.
5. Useful information can often be gained by observing the effect of an antimicrobial agent on the uptake kinetics of a radiolabelled nutrient, such as glucose, acetate, a fatty acid, an amino acid, a nucleic acid precursor, or phosphate. Changes in rate of incorporation after the addition of the drug are measured and compared with its effect on growth. A prompt interference with incorporation of a particular nutrient may provide a good clue to the primary site of action.
6. An antimicrobial compound which inhibits protein or nucleic acid synthesis in cells without interfering with membrane function or the biosynthesis of the immediate precursors of proteins and nucleic acids, or the generation and utilization of ATP, probably inhibits macromolecular synthesis directly. Because of the close interrelationship between protein and nucleic acid synthesis, indirect effects of the inhibition of one process on another process must be carefully distinguished. For example, inhibitors of the biosynthesis of RNA also block protein biosynthesis as the supply of messenger RNA (mRNA) is exhausted. Again, inhibitors of protein synthesis eventually arrest DNA synthesis because of the requirement for continued protein biosynthesis for the initiation of new cycles of DNA replication. A study of the kinetics of the inhibition of each macromolecular biosynthesis in intact cells is valuable since indirect inhibitions appear later than direct effects.
7. After the target biochemical system has been identified in intact cells, more detailed information is obtained with preparations of enzymes, nucleic acids and subcellular organelles. The antimicrobial compound is

tested for inhibitory activity against the suspected target reaction *in vitro*. There is a risk, however, of nonspecific drug effects *in vitro*, especially when the drug is added at high concentrations. Failure to inhibit the suspected target reaction *in vitro*, on the other hand, even with high drug concentrations, may not rule out inhibition of the same reaction in intact cells for several reasons.

(a) The drug may be metabolized by the host or the living micro-organism to an active, inhibitory derivative.
(b) The procedures involved in the purification of an enzyme may cause desensitization to the inhibitor by altering the conformation of the inhibitor binding site.
(c) The site of inhibition in the intact cell may be part of a highly integrated structure which is disrupted during the preparation of a cell-free system, again causing a loss of sensitivity to the inhibitor.

Enzyme or organelle preparations from drug-resistant mutants have been successfully used in identifying the site of attack, and examples of this approach are described in later chapters. Cloning procedures and recombinant DNA technology greatly facilitate the provision of suspected protein targets for *in vitro* evaluation.

Application of microarray expression and proteomic technologies in analyzing drug action

The acquisition of microbial genomic sequences and remarkable technological developments in molecular biology provide opportunities to profile the effects of antimicrobial drugs by investigating their effects on the expression of thousands of genes simultaneously. Although an antimicrobial drug may target a specific molecular receptor, its consequent effects on microbial metabolism and gene expression are not only pleiotropic, i.e. multiple, but may also be characteristic. The ability to assess the impact of drugs on the expression of many different genes simultaneously enables investigators to place compounds with closely similar primary sites of action into related sets. Gene transcription profiling of novel agents with unknown sites of action may therefore provide valuable clues as to their primary target receptors. One microarray technology involves the synthesis of short oligonucleotides in a high-density array directly on a solid surface, or 'chip'. The oligonucleotides are selected using total genomic DNA from the micro-organism to represent each open reading frame (orf). In this way many thousands of genes (or at least fragments of genes) can be arrayed. Messenger RNA extracted from cells cultured in the presence and absence of the drug under investigation is then hybridized to the immobilized oligonucleotides to reveal how the levels of expression of individual microbial genes are affected by the drug. There are numerous opportunities for experimental artefacts in analyzing microarray expression, and various controls are essential to ensure reproducible data. With this caveat in mind, patterns of gene expression can be obtained which are characteristic of specific modes of drug action, i.e. inhibition of protein or nucleic acid synthesis. Microarray experiments yield thousands of data points and the evaluation of such large amounts of information is a considerable challenge. Several different methods of data analysis are used, including hierarchical clustering, self-organizing maps, principal components analysis and vector algebra. It is worth restating that even with all this methodology, analysis of microarray expression is concerned with finding patterns of responses to the inhibitory actions of antimicrobial agents and is not currently capable of precisely defining the primary molecular target of drug action.

Similar comments apply to proteomic analysis of drug action. This method assesses the effects of drugs on the patterns of protein expression in microbial cells. Recently, proteomic analysis was applied to the effects of 30 antibacterial drugs from all the important types of agents on protein expression in *Bacillus subtilis*. Two-dimensional polyacrylamide electrophoresis was used to separate radiolabelled cytoplasmic proteins. Each of the 30 drugs produced complex but reproducible and characteristic patterns of protein expression. Combinations of microarray RNA expression and proteomic technologies should eventually offer a highly refined approach to the characterization of drug action and expedite the ultimate definition of the primary molecular targets.

1.4.3 The study of the interactions between antimicrobial agents and their molecular target

Early mode of action studies concentrated on revealing the biochemical processes and pathways inhibited by antimicrobial drugs. However, scientists are no longer satisfied with this level of explanation alone and aspire to define drug action in molecular terms, i.e. the details of the specific interactions between drugs and their target sites. In order to achieve this level of understanding of drug action, techniques such as X-ray crystallography and nuclear magnetic resonance spectroscopy (NMR) are used to generate visual images of the molecular interactions between drug and macromolecule. Recombinant DNA technology enables the role of specific amino acids or nucleotide residues in macromolecule–drug interactions to be defined. The structural elucidation of supramolecular organized structures, such as membranes and ribosomes, proved to be a more formidable undertaking, largely because of the difficulty in obtaining diffracting crystals of these structures for X-ray analysis. Nevertheless, in the past few years even the structure of bacterial ribosomes has been revealed by X-ray crystallography, together with remarkable details of their interactions with drugs of major clinical importance which inhibit protein biosynthesis (Chapter 5).

1.4.4 Pharmacological biochemistry

Effective antimicrobial drugs possess a combination of advantageous properties: potency and selectivity at the target site, good absorption from the site of administration, appropriate distribution within the body of the infected host, adequate persistence in the tissues and absence of significant toxicity to the patient. Each of these attributes may require distinct molecular characteristics. For optimum activity, all these characteristics must be combined in the same molecule. The absorption, distribution, metabolism, excretion and toxicity of drugs are therefore essential subjects for investigation. Activity requires an inhibitory concentration of the drug at the target site which must be sustained long enough to allow the body's defences to contribute to the defeat of the infection. The concentration is determined by the rates of absorption and excretion and also by the metabolism of the drug in the tissues of the host. The extent of binding of the drug to host proteins can also be important. While extensive binding to plasma proteins can increase drug persistence in the body, it may also reduce effectiveness because the activity of drugs depends on the concentration of free (unbound) compound in the immediate environment of the infecting micro-organism. The methods for studying such factors using modern analytical techniques are well established. The data can help to explain species differences in the therapeutic activities of drugs and provide a sound basis for recommendations on the size and frequency of doses for treating patients. The study of the pharmacological biochemistry of drugs is a highly specialized field and is beyond the scope of this book; only passing mention is therefore made of the pharmacological factors that influence the activities of antimicrobial drugs.

1.4.5 Selectivity of action of antimicrobial agents

Safe and effective antimicrobial drugs are by definition highly selective in their action on the infecting pathogens. In some cases the molecular targets inhibited by drugs are specific to the microbial cell. In other cases drugs may act on biochemical mechanisms that are common to both microbe and host. Possible explanations for selectivity of action in the latter situation can include sufficient structural differences between the microbial and host enzymes catalyzing the same reaction to permit a selective attack on the microbial enzyme, specificity due to selective concentration of the drug within the microbial cell, and finally, differences between the rates of turnover of target molecules in the pathogen and its host that provide a basis for selectivity of action (see eflornithine, Chapter 6).

1.4.6 Biochemistry of microbial resistance

The effectiveness of antimicrobial drugs frequently declines after sustained use, owing to the emergence of drug-resistant organisms. This enormously important problem has been studied in great depth by microbiological, biochemical and molecular genetic methods. Such studies have revealed the genetic basis for the emergence of drug-resistant bacteria, fungi and viruses (Chapter 8) and defined the biochemical mechanisms of resistance (Chapter 9). The mechanism of some forms of resistance in eukaryotic pathogens, such as chloroquine resistance in the malarial parasite, is proving more difficult to define but is nevertheless the subject of much research attention because of the re-emergence of malaria across much of the tropical and subtropical regions of the world.

1.5 Current trends in the discovery of antimicrobial drugs

The relentless increase in drug resistance among bacteria, fungi, viruses and protozoal pathogens provides an urgent stimulus for the discovery of new antimicrobial drugs. It is hoped that several developments in research technologies during the past decade will facilitate the discovery process; these are briefly outlined below.

1.5.1 Bioinformatics and genomics

Many of the antimicrobial drugs in current clinical use were discovered empirically by screening against cultures of micro-organisms or directly against model infections in experimental animals. Explanations of the biochemical and molecular basis of drug action often emerged only after years of use in patients. In contrast, the modern approach to drug discovery is usually driven by the perceived molecular target. Now that at least 70 complete bacterial genomic sequences are available, in addition to sequences from some viruses, pathogenic yeasts and the malarial parasite, *Plasmodium falciparum*, the enormous and ever-expanding wealth of genomic information from both micro-organisms and their mammalian hosts provides many opportunities to identify potential molecular targets which are either pathogen-specific or sufficiently different in sequence from the mammalian counterparts to offer the possibility of drug selectivity. The computer-based technology of bioinformatics is used to search micro-organisms for proteins with highly conserved sequences which suggest functions that are essential for viability. Validation of such proteins as drug targets is then obtained by targeted gene knockout experiments in pathogens grown *in vitro* and in infected animals. In the search for potential broad-spectrum antibacterial drug activity, target proteins are sought that have a high degree of sequence identity across the major pathogens, both Gram-positive and Gram-negative.

1.5.2 High-throughput screening versus targeted screening

After an attractive molecular target has been selected by bioinformatic 'data mining' and a microbiological demonstration of essentiality for viability has been carried out, sufficient protein is produced by gene cloning and expression technologies to allow screening to begin. Advances in laboratory robotic procedures have made it possible to rapidly screen extremely large numbers of compounds. Whereas 20 years ago screening perhaps 100 or so compounds per week against an enzyme target might have been considered satisfactory, screening a million compounds per week is now not unusual. The enormous quantities of data generated by such high-throughput screens created new challenges in evaluating and filtering compounds of interest for the succeeding stages of screening cascades. The provision of chemicals in such vast numbers also places considerable demands on compound collections and on synthetic chemistry. Fortunately the miniaturized techniques of combinatorial chemistry are capable of synthesizing prodigious numbers of compounds.

For some scientists, however, the intellectual appeal of high-throughput screening is limited and its productivity in delivering novel effective drugs has been questioned. An alternative approach is that of targeted screening, based upon an appreciation of the structure and function of the target protein. The ability of genetic engineering and protein production facilities to provide substantial quantities of previously inaccessible proteins has expanded the opportunities for the determination of three-dimensional structures by X-ray crystallography and NMR. Visualization of protein structure on the computer monitor screen enables chemists to design potential inhibitors *in silico* which can then be synthesized and evaluated in targeted biochemical screens. At this stage, the numbers of compounds may be quite small but they are soon expanded as promising compounds are further modified to optimize potency of inhibition, activity and safety *in vivo*. In all likelihood both high-throughput and targeted screening approaches will be employed for the foreseeable future in the search for novel, effective antimicrobial drugs.

1.6 Scope and layout of the book

In this book we have sought to provide well-established evidence for the biochemical actions and molecular targets of many of the best-known agents used in medicine. Although much of the content is devoted to antibacterial drugs, which constitute by far the largest group of antimicrobial agents in use today, we have also brought together information on the biochemical activities of commonly used antifungal, antiprotozoal and antiviral drugs. One chapter is devoted to the means by which antimicrobial compounds enter and leave their target cells and their relevance to the intrinsic resistance of micro-organisms to drugs. The last two chapters consider the genetic and biochemical basis of acquired drug resistance, respectively.

Wherever possible we have grouped drugs according to their types of biochemical action rather than by their therapeutic targets. However, one chapter brings together several drugs with other and unusual modes of action.

Further reading

Bandow, J. E. *et al.* (2003) Proteomic approach to understanding antibiotic action. *Antimicrob. Agents Chemother.* **47**, 948.

Greenwood, D. (1995) *Antimicrobial Chemotherapy*. 3rd Edn. Oxford University Press, Oxford.

Le Fanu, J. (1999) *The Rise and Fall of Modern Medicine*. Little, Brown & Co., New York.

Sneader, W. (1985) *Drug Discovery: The Evolution of Modern Medicines*. John Wiley & Sons, New York.

Volker C. and Brown, J. R. (2002). Bioinformatics and the discovery of novel anti-microbial agents. *Curr. Drug Targets*. **2**, 279.

Chapter two

Vulnerable shields—the cell walls of bacteria and fungi

2.1 Functions of the cell wall

In the search for differences between microbial pathogens and animal cells that could provide the basis for selective antimicrobial attack, one evident distinction lies in their general structure. The animal cell is relatively large and has a complex organization; its biochemical processes are compartmentalized and different functions are served by the nucleus with its surrounding membrane, by the mitochondria and by various other organelles. The cytoplasmic membrane is thin and lacks rigidity. The cell exists in an environment controlled in temperature and osmolariy in mammals and birds. It is constantly supplied with nutrients from the extracellular fluid. Bacteria and fungi live in variable and often hostile environments and they must be able to withstand considerable changes in external osmolarity. Some micro-organisms have relatively high concentrations of low molecular weight solutes in their cytoplasm. Such cells suspended in water or in dilute solutions develop a high internal osmotic pressure. This would inevitably disrupt the cytoplasmic membrane unless it were provided with a tough, elastic outer coat. This coat is the cell wall, a characteristic of bacteria and fungi which is entirely lacking in animal cells. It has a protective function but at the same time it is vulnerable to attack, and a number of antibacterial and antifungal drugs owe their action to their ability to disturb the processes by which the walls are synthesized. Since there is no parallel biosynthetic mechanism in animal cells, substances affecting this process may be highly selective in their antimicrobial action.

The term 'wall' will be used to describe all the cell covering which lies outside the cytoplasmic membrane. The structures of the walls of bacteria and fungi are very different from each other, as are the biosynthetic processes involved in their elaboration. This results in susceptibility to quite distinct antimicrobial agents.

2.2 Structure of the bacterial wall

The structure of the bacterial wall not only differs markedly from that of fungi but also varies considerably from one species to another. It nevertheless follows general patterns which are related to the broad morphological classification of bacteria. Classically this has been based on the responses towards the Gram stain, but the well-tried division into Gram-positive and Gram-negative types has a significance far beyond that of an empirical staining reaction. The most evi dent differences are worth recalling.

Many Gram-negative bacteria are highly adaptable organisms which can use inorganic nitrogen compounds, mineral salts and a simple carbon source for the synthesis of their whole structure. Their cytoplasm has a relatively low osmolarity. Typical Gram-positive cocci or bacilli tend to be more exacting in their nutritional needs. They are usually cultivated on rich undefined broths or on fairly elaborate synthetic media. In their cytoplasm, Gram-positive bacteria concentrate amino acids, nucleotides and other metabolites of low molecular weight and consequently have a high internal osmolarity. However, not all bacteria fit this neat division. The Gram-negative cocci, the rickettsias, the chlamydias and the spirochetes, for example, are all Gram-negative bacteria with exacting growth requirements. The mycoplasmas lack a rigid wall structure and although technically Gram-negative, they are best treated as a separate group lying outside the usual Gram stain classification.

For many years the bacterial wall was considered to be a rigid structure, largely because when bacteria are disrupted, the isolated walls retain the shape of the intact organisms. More recent evidence, however, shows that this concept of rigidity must be revised. The peptidoglycan sacculus (see later discussion) of the bacterial wall can expand or contract in response to changes in the ionic strength or the pH of the external environment. This responsive flexibility is a property of the wall itself and can even be seen by the unaided eye when salt solutions are added to quantities of walls pelleted by centrifugation. When intact bacteria are subjected to osmotic stress, water moves through the wall and membrane into the cytoplasm. The consequent swelling of the cell, bounded by the membrane, is accommodated to some extent by the limited elasticity of the wall, although even stretchable structures break wher sufficiently stressed. The wall breaks and the cell then bursts as a result of the turgor pressure on the thin cytoplasmic membrane. It is important to realize that during cell growth and proliferation, bacterial cell walls are highly dynamic structures, continually undergoing biosynthesis, extension and remodeling. It is this dynamic character which renders bacteria susceptible to antibiotics which attack the biosynthesis and the integrity of cell walls.

Most of the work on wall structure has been done with Gram-positive cocci and bacilli and with enteric bacteria and other Gram-negative rods. The extent to which the structural generalizations apply to groups outside these classes is uncertain.

2.2.1 The Gram-positive wall

The basic structure of the cell walls of Gram-positive bacteria is relatively simple (Figure 2.1), although there are many differences of detail across the species. The wall which lies outside the cytoplasmic membrane is usually beween 15 and 50 nm thick. Bacteria can be broken by shaking with small glass beads and the walls separated from cytoplasmic material by washing and differential centrifugation. In electron micrographs these wall preparations resemble empty envelopes torn in places where the cytoplasmic contents were released. The major part of the Gram-positive wall is a large polymer consisting of two covalently linked components. One of these components, forming at least 50% of the wall mass, is peptidoglycan (sometimes referred to as murein or mucopeptide). Its cross-linked structure provides a tough, fibrous fabric that gives strength and shape to the cell and enables it to withstand a high internal osmotic pressure. The amount of peptidoglycan in the wall shows that it covers the cell in a multilayered fashion, with cross-linking both within and between the layers. Attached to the peptidoglycan is an acidic polymer, accounting for 30–40% of the wall mass, which differs from species to species. Often this is a teichoic acid—a substituted poly(D-ribitol 5-phosphate) (see Figure 2.8)—or a substituted glycerol 3-phosphate (lipoteichoic acid). In some bacteria teichoic acid is replaced by poly(*N*-acetylglucosamine 1-phosphate) or teichuronic acid (a polymer containing uronic acid and *N*-acetylhexosamine units). Bacteria that normally incorporate teichoic acid in their walls can switch to teichuronic acid under conditions of phosphate limitation. The acidic character of the polymer attached to the peptidoglycan ensures that the cell surface is strongly polar and carries a negative charge. This may influence the passage of ions, particularly Mg^{2+} and possibly ionized drugs, into the cell. The teichoic acid or other acidic polymer is readily solubilized and released from the insoluble peptidoglycan by hydrolysis in cold acid or alkali. The nature of the linkage is described later.

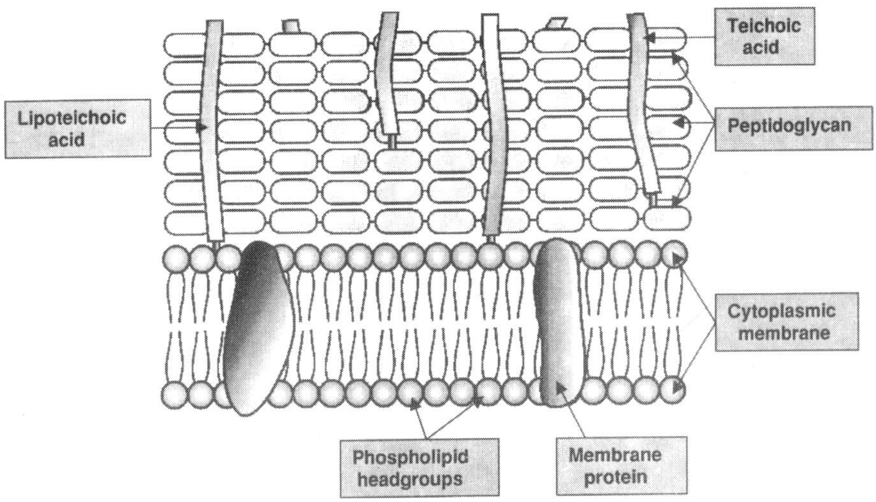

FIGURE 2.1 The arrangement of the cell envelope of Gram-positive bacteria. Note that the term 'cell envelope' includes both the cytoplasmic membrane and the outer layers of the cell. The components are not drawn to scale. (This diagram was kindly provided by Philip Kerkhoff.)

Other components of the Gram-positive wall vary widely from species to species. Protein is often present to the extent of 5–10%, and protein A of *Staphylococcus aureus* is apparently linked covalently to peptidoglycan. Proteins and polysaccharides frequently occur in the outermost layers and provide the main source of the antigenic properties of these bacteria. Mycobacteria and a few related genera differ from other Gram-positive bacteria in having large amounts of complex lipids in their wall structure. The unique features of the mycobacterial cell wall are described later in this chapter.

2.2.2 The Gram-negative wall

The Gram-negative wall is far more complex. Wide-ranging studies of its structure have been concentrated on the Enterobacteriaceae and on *Escherichia coli* in particular. The diagram in Figure 2.2 illustrates the general arrangement of the components of the Gram-negative cell envelope, which includes the cytoplasmic membrane as well as the cell wall. When cells of *Escherichia coli* are fixed, stained with suitable metal salts, sectioned and examined by electron microscopy, the cytoplasmic membrane is readily identified by its 'sandwich' appearance of two electron-dense layers separated by a lighter space. The clear layer immediately outside the cytoplasmic membrane has been described as the periplasmic space. However, techniques in electron microscopy such as freeze-etching and freeze-substitution reveal that a rich, dense periplasm occupies the periplasmic 'space', containing a wealth of biochemicals, including enzymes, transport proteins, secreted materials, components of peptidoglycan and the bacterial outer membrane (see later discussion). The electron-dense layer, about 2 nm thick, immediately outside the periplasm represents the peptidoglycan component of the wall. It is much thinner than in Gram-positive bacteria and may constitute only 5 to 10% of the wall mass. Even so, it contributes substantially to wall strength. Cells rapidly lyse when treated with lysozyme, an enzyme which specifically degrades peptidoglycan. In *Escherichia coli* the peptidoglycan is covalently linked to a lipoprotein which probably projects into the outer regions of the wall. The outer regions of the Gram-negative cell wall have been the most difficult to characterize. The various

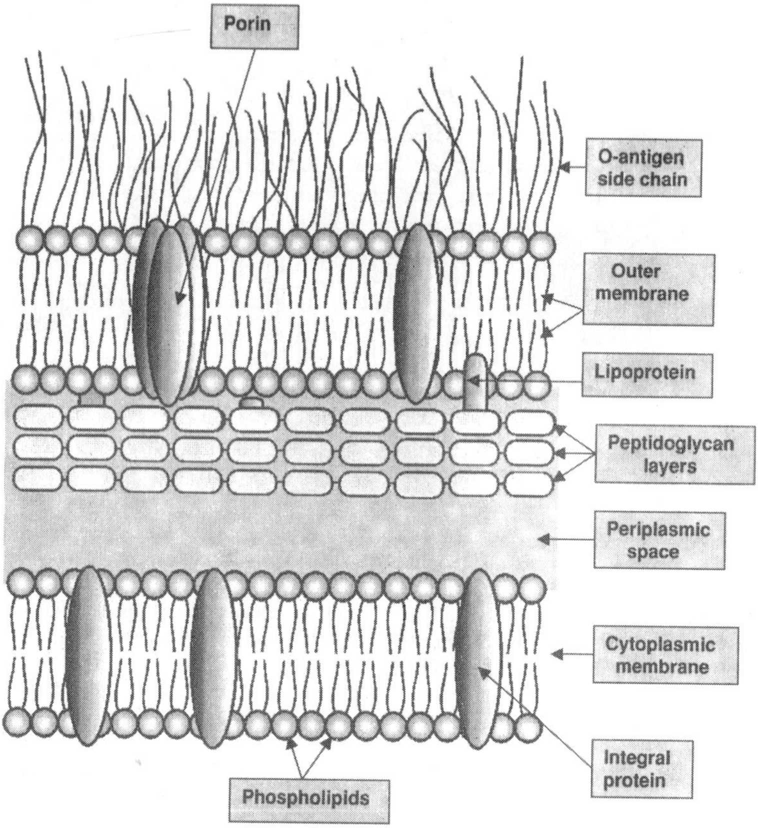

FIGURE 2.2 The arrangement of the various layers of the cell envelope of Gram-negative bacteria. The components are not drawn to scale. (This diagram was kindly provided by Philip Kerhoff.)

components together form a structure 6–10 nm thick, called the outer membrane. Like the cytoplasmic membrane, it is basically a lipid bilayer (giving rise to the two outermost electron-dense bands), hydrophobic in the interior with hydrophilic groups at the outer surfaces. It also has protein components which penetrate the layer partly or completely and form the membrane 'mosaic'.

Despite these broad structural similarities, the outer membrane differs widely in composition and function from the cytoplasmic membrane. Its main constituents are a lipopolysaccharide, phospholipids, fatty acids and proteins. The phospholipids, mainly phosphatidylethanolamine and phosphatidylglycerol, resemble those in the cytoplasmic membrane. The structure of the lipopolysaccharide is complex and varies considerably from one bacterial strain to another. The molecule has three parts (Figure 2.3). The core is built from 3-deoxy-D-*manno*-octulosonic acid (KDO), hexoses, heptoses, ethanolamine and phosphoric acid as structural components. The three KDO residues contribute a structural unit which strongly binds the divalent ions of magnesium and calcium, an important feature that stabilizes the membrane. Removal of these ions by chelating agents leads to release of some of the lipopolysaccharide into the medium; at

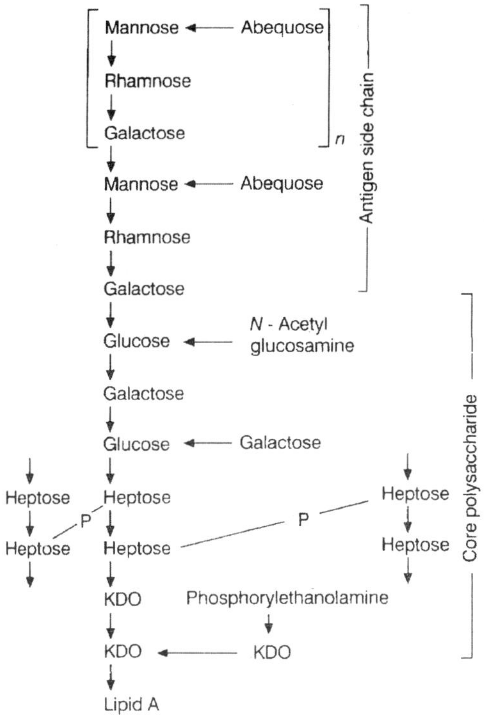

FIGURE 2.3 Structure of the lipopolysaccharide of the cell envelope of *Salmonella typhimurium*. The diagram has been simplified by omitting the configuration of the glycosidic linkages and omitting the *O*-acetyl groups from the abequose units. KDO: 3-deoxy-D-*manno*-octulosonic acid. Lipid A consists of a β-1,6-linked diglucosamine residue to which lauric, myristic, palmitic and 3-D(-)-hydroxymyristic acids are bound. The heptose residues of three lipopolysaccharide polymers are shown linked by phosphate diester bridges. Although there are considerable structural variations in the antigen side chains among *Salmonella* species, the core polysaccharide and lipid A are probably common to all wild-type salmonellae. The core structure in *Escherichia coli* is more variable.

the same time, the membrane becomes permeable to compounds that would otherwise be excluded. The core polysaccharide is linked to the antigenic side chain, a polysaccharide which can vary greatly from one strain to another even within the same bacterial species. Usually it consists of about 30 sugar units, although these can vary in both number and structure. It forms the outermost layer of the cell and is the main source of its antigenic characteristics. At the opposite end, the core of the lipopolysaccharide is attached to a moiety known as lipid A which can be hydrolyzed to glucosamine, long-chain fatty acids, phosphate and ethanolamine. The fatty acid chains of lipid A, along with those of the phospholipids, align themselves to form the hydrophobic interior of the membrane. The outer membrane is therefore asymmetric, with lipopolysaccharide exclusively on the outer surface and phospholipid mainly on the inner surface.

The most abundant proteins of the outer membrane in *Escherichia coli* are the porin proteins and lipoprotein. Electron microscopy of spheroplasts lacking peptidoglycan reveals triplets of indentations in the membrane surface, each 2 nm in diameter and 3 nm apart, through which the stain used in the preparation readily penetrates. This is interpreted as showing that the porin protein molecules stretch across the membrane in groups of three, enclosing pores through which water and small molecules can diffuse. The size of the pores explains the selective permeability of the Gram-negative outer membrane; they freely allow the passage of hydrophilic molecules up to a maximum molecular weight of 600–700. Larger flexible molecules may also diffuse through the pores, although with more difficulty. Artificial vesicles can be made with outer membrane lipids. Without protein, these are impermeable to solutes, but when porins are incorporated, they show permeability characteristics similar to those of the outer membrane itself. The role of porins in influencing the penetration of antibacterial drugs into Gram-negative bacteria is explored in Chapter 7.

Lipoprotein is another major component of the outer membrane proteins. About one-third is linked to peptidoglycan and the remaining two-thirds are unattached but form part of the membrane. The nature of the attachment of lipoprotein to the side chains of peptidoglycan is discussed later. About one in twelve of the peptide side chains is substituted in this way. This arrangement anchors the outer membrane to the peptidoglycan layer. The fatty acid chains of the lipoprotein presumably align themselves in the hydrophobic inner layer of the outer membrane and the protein moiety may possibly associate with matrix protein, reinforcing the pore structure.

Many other proteins with specialized functions have been identified in the outer membrane. Some of these are transport proteins that allow access to molecules such as vitamin B_{12} or nucleosides which are too large to penetrate the pores of the membrane. Outer membrane proteins that contribute to the function of multidrug efflux pumps are described in Chapter 7.

2.3 Structure and biosynthesis of peptidoglycan

The structure and biosynthesis of peptidoglycan have special significance relative to the action of a number of important antibacterials and have been studied extensively. The biosynthesis of peptidoglycan was first worked out with *Staphylococcus aureus*. Although bacteria show many variations in peptidoglycan structure, the biosynthetic sequence in *Staphylococcus aureus* illustrates the general features of the process. In this description the enzymes involved will be referred to, where appropriate, by their biochemical names and also by the more recent popular abbreviations derived from the genetic nomenclature. The biosynthetic sequence may be conveniently divided into four stages.

2.3.1 Stage 1: Synthesis of UDP-*N*-acetylmuramic acid

The biosynthesis starts in the cytoplasm with two products from the normal metabolic pool, *N*-acetylglucosamine 1-phosphate and UTP (Figure 2.4). UDP-*N*-acetylglucosamine (I) formation is catalyzed by *N*-acetyl-1-phosphate glucosamine uridyl transferase (GlmU) with the elimination of pyrophosphate. This nucleotide reacts with phosphoenol pyruvate catalyzed by UDP-*N*-acetylglucosamine enolpyruvyl transferase (MurA) to give the corresponding 3-enolpyruvyl ether (II). The pyruvyl group is then converted to lactyl by a reductase (MurB) that requires both flavin NAD and NADPH as cofactors, the product being UDP-*N*-acetylmuramic acid (III, UDPMurNAc). Muramic acid (3-*O*-D-lactyl-D-glucosamine) is a distinctive amino sugar derivative found only in the peptidoglycan of cell walls.

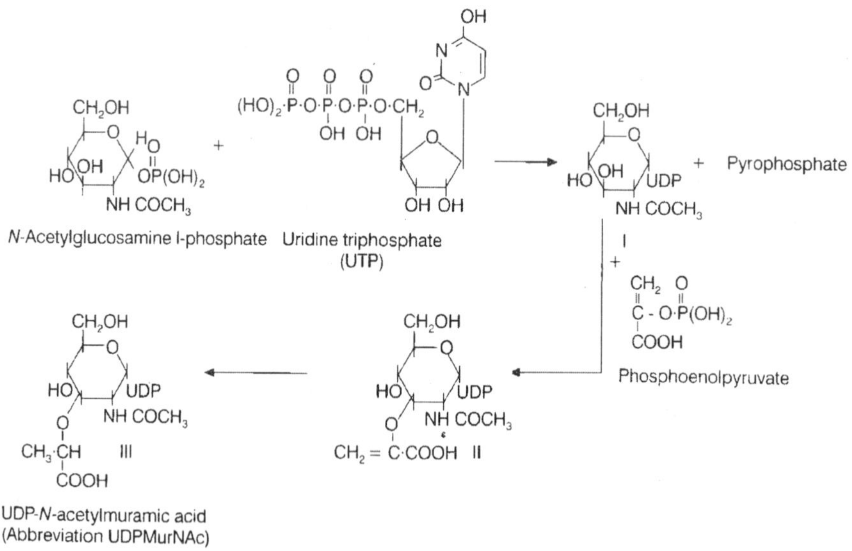

FIGURE 2.4 Peptidoglycan synthesis in *Staphylococcus aureus*. Stage 1: formation of UDP-*N*-acetylmuramic acid.

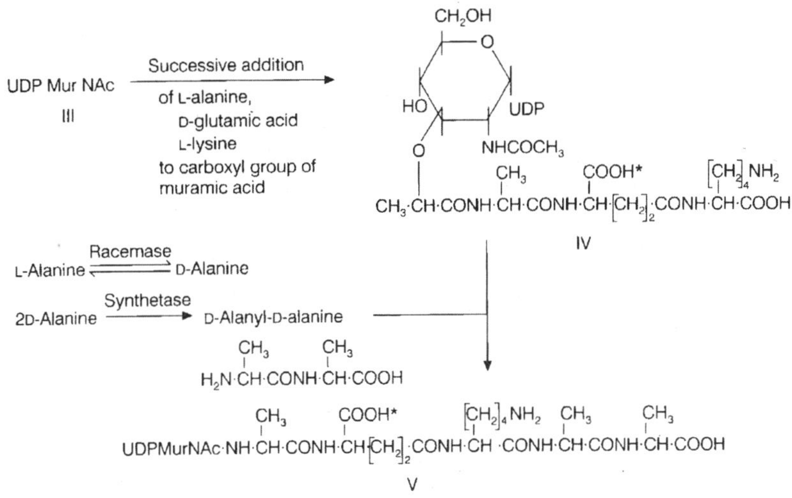

UDP-*N*-acetylmuramyl pentapeptide

FIGURE 2.5 Peptidoglycan synthesis in *Staphylococcus aureus*. Stage 2: formation of UDP-*N*-acetylmuramyl pentapeptide. Addition of each amino acid and the final dipeptide requires ATP and a specific enzyme. L-Lysine is added to the γ-carboxyl group of D-glutamic acid; the α-carboxyl group (marked *) is amidated at a later stage in the biosynthesis.

2.3.2 Stage 2: Building the pentapeptide side chain

Five amino acid residues are next added to the carboxyl group of the muramic acid nucleotide (Figure 2.5). Each step requires ATP and a specific amino acid ligase. L-Alanine is added first by MurC. The next two residues added are D-glutamic acid catalyzed by MurD and then either L lysine or *meso* diaminopimelic acid by amino acid-specific forms of MurE. The incorporation of either L-lysine or *meso*-diaminopimelic acid into the pentapeptide side chain is characteristic of individual bacterial species. These latter amino acids are attached via their α-amino groups to the γ-carboxyl group of the glutamic acid. In *Staphylococus aureus* and *Streptococcus pneumoniae*, but not in other bacteria, the α-carboxyl group of the glutamic acid is amidated at a later stage in the biosynthesis; this amino acid residue is sometimes referred to as D-isoglutamine. The biosynthesis of the pentapeptide is completed by addition, not of an amino acid, but of a dipeptide, D-alanyl-D-alanine, which is synthesized separately. A racemase acting on L-alanine converts it to D-alanine, and a ligase joins two molecules of D-alanine to give the dipeptide. The linkage of D-alanyl-D-alanine to the tripeptide chain is catalyzed by MurF. The completed UDP-*N*-acetylmuramyl intermediate (V) with its pendant peptide group will be referred to as the 'nucleotide pentapeptide'.

The three-dimensional structures of MurC, MurD, MurE and MurF have all been solved by X-ray crystallography and are generally very similar. Since these enzymes are unique to bacteria, detailed knowledge of the structures may eventually lead to the design of highly specific antibacterial drugs.

2.3.3 Stage 3: Membrane-bound reactions leading to a linear peptidoglycan polymer

The biosynthesis up to this point is cytoplasmic, while the succeeding steps occur on membrane structures. The first membrane-associated step involves the formation of a pyrophosphate link, catalyzed by

UDP-*N*-acetylmuramyl pentapeptide phosphotransferase (MraY), between the nucleotide pentapeptide and undecaprenyl phosphate (the phosphate ester of a C_{55} isoprenoid alcohol), which is a component of the cytoplasmic membrane, to form a complex referred to as lipid I. In this reaction UMP is released and becomes available for reconversion to UTP, which is needed in the first step of peptidoglycan biosynthesis (Figure 2.6). All subsequent reactions occurring while the intermediates are linked to undecaprenyl phosphate take place without release from the membrane. An essential step in this membrane-bound reaction sequence is the addition of a second hexosamine residue through a typical glycosidation by UDP-*N*-acetylglucosamine catalyzed by a glycosyl transferase (MurG) (Figure 2.6). The modified disaccharide known as lipid II is formed by a 1,4-β linkage with liberation of UDP. The involvement of undecaprenyl phosphate is not unique to peptidoglycan biosynthesis. It is also concerned in the biosynthesis of the polysaccharide chain in the *O*-antigen produced by *Salmonella typhimurium* and in the formation of the polysaccharide elements of the lipopolysaccharides of Gram-negative bacteria; in Gram-positive bacteria it fulfils a similar role in the biosynthesis of teichoic acid or polysaccharides of the wall. The structure of MurG has also been solved and active sites identified for inhibitor design studies.

At about this point in the biosynthesis of *Staphylococcus aureus* peptidoglycan and in many other Gram-positive bacteria, an extending group is added to the ε-amino group of the lysine unit in the nucleotide pentapeptide. Glycine and a glycine-specific transfer RNA (tRNA) are involved in this process during which a pentaglycine group is added. This reaction, which is not found in Gram-negative bacteria, is unlike the tRNA reactions in protein biosynthesis because ribosomes are not involved; the five glycine units are added successively to the lysine from the nitrogen end (the reverse direction of protein biosynthesis). The resultant product (VIII, Figure 2.6) with 10 amino acid units is referred to as the disaccharide decapeptide and retains a free terminal amino group. In the biosynthesis of peptidoglycans in certain other bacterial species, for example in *Escherichia coli*, in which no extending group is added, the later reactions involve the ε-amino group of *meso*-diaminopimelic acid (or equivalent diamino acid) instead of the terminal amino group of glycine. During the membrane-bound stage in the biosynthesis of *Staphylococcus aureus* peptidoglycan, the carboxyl group of D-glutamic acid is amidated by a reaction with ammonia and ATP.

The disaccharide decapeptide (VIII) is now attached to an 'acceptor', usually regarded as the growing linear polymer chain. In this reaction the disaccharide with its decapeptide side chain forms a β-linkage from the 1 position of the *N*-acetylmuramic acid residue to the 4-hydroxyl group of the terminal *N*-acetylglucosamine residue in the growing polysaccharide chain. Because this reaction occurs outside the cytoplasmic membrane, the disaccharide-decapeptide linked to the undecaprenyl phosphate (lipid II) first moves across the membrane to gain access to the acceptor on the external face of the membrane. The released undecaprenyl pyrophosphate is reconverted by a specific pyrophosphatase to the corresponding phosphate, ready for another cycle of the membrane-bound part of the synthesis. The extension of the glycan chains thus occurs by successive addition of disaccharide units catalyzed by glycosyl transferases.

2.3.4 Stage 4. Cross-linking

The linear peptidoglycan (IX, Figure 2.6) formed in stage 3 contains many polar groups which make it soluble in water. It lacks mechanical strength and toughness. These attributes are introduced in the final stage of biosynthesis by cross-linking, a process well known in the plastics industry for producing similar results in synthetic linear polymers. The mechanism involved in cross-linking peptidoglycan is a transpeptidation reaction requiring no external supply of ATP or similar compounds. In *Staphylococcus aureus*, the transpeptidation occurs between the terminal amino group of the pentaglycine, side chain and the peptide amino group of the terminal D-alanine residue of another peptide side chain; D-alanine is eliminated and a peptide bond formed (Figure 2.7). In *Staphylococcus aureus* peptidoglycan, the cross-linking is quite extensive and up to 10 peptide side chains may be bound together by bridging groups. Since the linear polymers themselves are very large, it is likely that the whole of the peptidoglycan in a Gram-positive bacterium is made up of

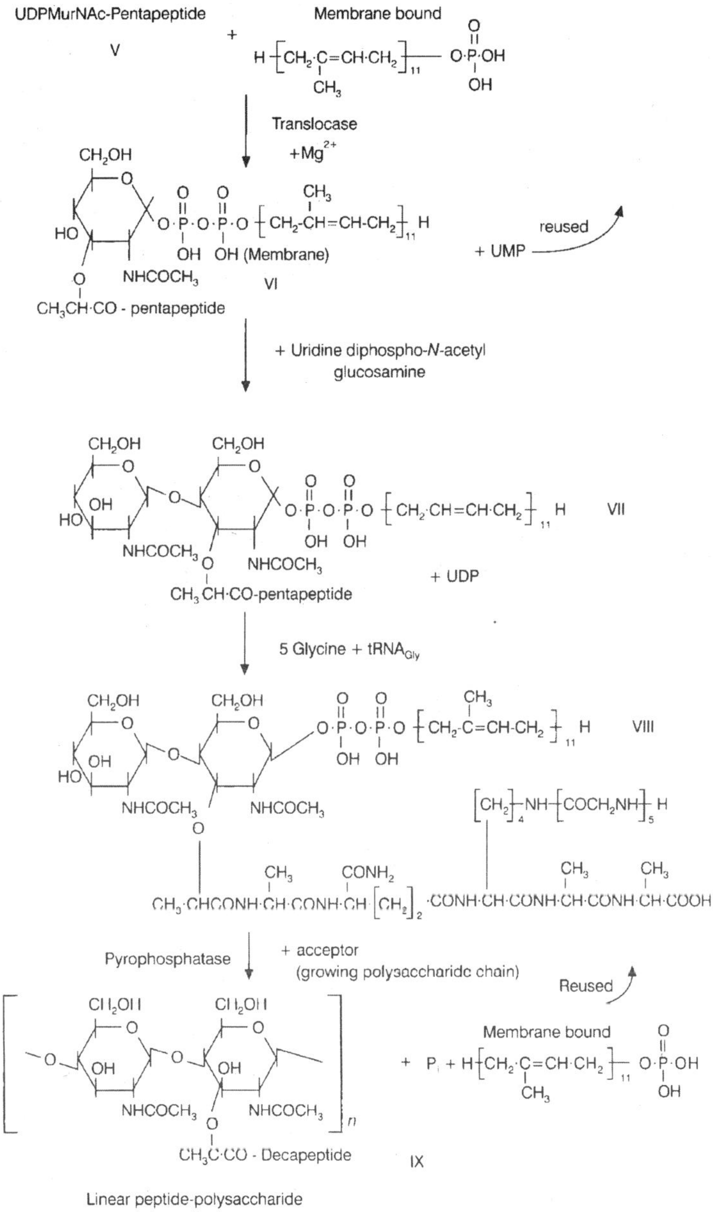

FIGURE 2.6 Peptidoglycan synthesis in *Staphylococcus aureus*. Stage 3: formation of the linear peptidoglycan. The structure of the decapeptide side chain is shown in VIII.

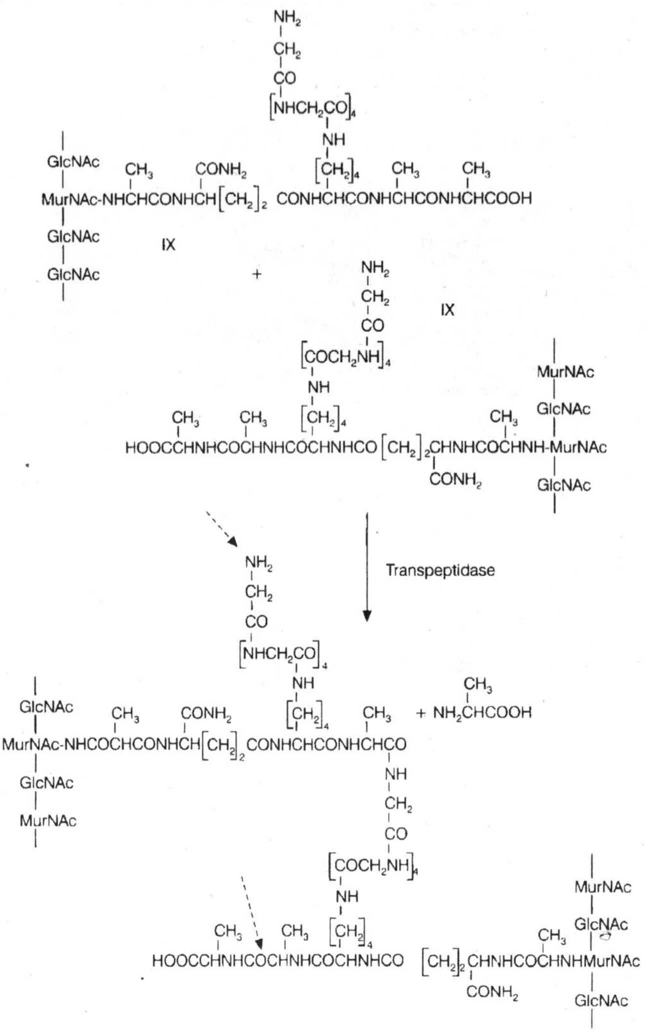

FIGURE 2.7 Peptidoglycan synthesis in *Staphylococcus aureus*. Stage 4: cross-linking of two linear peptidoglycan chains. The linear polymers have the structure IX (Figure 2.6) GlcNac: *N*-acetylglucosaminyl residue. The dashed arrows show points at which further cross-links may be formed with other polymer chains. MurNAc: *N*-acetylmuramyl residue.

units covalently bound together. This gigantic bag-shaped molecule has been called a sacculus. There is also a mechanism for constantly breaking it down and reforming it to allow cell growth and division. Peptidoglycan hydrolases, which hydrolyze the polysaccharide chains of peptidoglycan and others attacking the peptide cross-links, exert this essential catabolic activity during cell growth.

2.3.5 Penicillin-binding proteins

The membrane-bound enzymes involved in linking the disaccharide deca- or pentapeptide to the growing linear peptidoglycan and the subsequent cross-linking reaction are referred to as penicillin-binding proteins or PBPs (Table 2.1). The PBPs are regarded as the specific targets for penicillin and the other β-lactam an-

TABLE 2.1 Properties of penicillin-binding proteins of *Escherichia coli*

Protein no.	*Molecular mass (Kilodaltons)*	*Enzyme activities*	*Function*
1a	91	Transpeptidase Transglycosylase	Peptidoglycan cross-linking
1b	91	Transpeptidase Transglycosylase	Peptidoglycan cross-linking
2	66	Transpeptidase	Peptidoglycan cross-linking
3	60	Transpeptidase	Peptidoglycan cross-linking
4	49	DD-carboxypeptidase	Limitation of peptidoglycan cross-linking
5	41	DD-carboxypeptidase	Limitation of peptidoglycan cross-linking
6	40	DD-carboxypeptidase	Limitation of peptidoglycan cross-linking

tibiotics. As we shall see, the covalent reaction between β-lactam antibiotics and the PBPs, which inactivates their transpeptidase function but not the transglycosylase activity, is central to the antibacterial activity of these drugs. PBPs vary from species to species in number, size, amount and affinity for β-lactams antibiotics. The PBPs fall into two major groups of high (≥ 60 kDa) and low (≤ 49 kDa) molecular mass, respectively. The PBPs with a high molecular mass are essentially two-domain proteins classed as A or B. In both classes the C-terminal domain is responsible for transpeptidation and is the target for penicillin binding and β-lactam action. Class A proteins also catalyze the transglycosylation reactions at the N-terminal domains. PBPs 1a and 1b of *Escherichia coli* exemplify this bifunctional type. The monofunctional class B proteins lack transglycolase activity. Monofunctional glycosyl transferases have been identified in both Gram-positive and Gram-negative bacteria, although not all glycosyl transferases appear to be essential to bacterial viability. In *Escherichia coli*, PBPs 1a and 1b provide the key enzyme activities involved in peptidoglycan synthesis. The synthetic role of PBP2 is specifically involved in cellular elongation and that of PBP3 with the formation of the cell septum during cell division. The low molecular mass PBPs, which include PBPs 4, 5 and 6 in *Escherichia coli*, are monofunctional DD-carboxypeptidases that catalyze transfer reactions from D-alanyl-D-alanine terminated peptides. Although these PBPs are also inactivated by β-lactams, this may not be central to their antibiotic action. Nevertheless, carboxypeptidases of this type are convenient to purify and have been widely used as models for the nature of the interaction between PBPs and penicillin. The most widely studied enzymes are the extracellular DD-carboxypeptidases produced by *Streptomyces* species and carboxypeptidases solubilized from the membranes of *Escherichia coli* and *Bacillus stearothermophilus*. The *Streptomyces* enzymes display some transpeptidase activity besides their high carboxypeptidase activity.

In addition to the seven 'classic' PBPs listed in Table 2.1, a further five have been added to the collection: PBP1c, PBP7, DacD, AmpC and AmpH. An extensive study of deletion mutants reveals that only PBPs 2 and 3 plus PBP1a or 1b are essential for the growth and division of rod-shaped bacteria under laboratory conditions. However, combinations of the activities of the other PBPs may be necessary for growth and viability in more demanding conditions, for example, in an infected host. The possible significance of the 'new' five PBPs in relation to the antibacterial action of β-lactams remains to be explored.

2.3.6 Variations in peptidoglycan structure

Many variations are found in peptidoglycan structure between one species of bacteria and another or even between strains of the same species and only a general account is possible here. All peptidoglycans have the same glycan chain as in *Staphylococcus aureus* except that the glucosamine residues are sometimes *N*-acylated with a group other than acetyl. *O*-Acetylation of glucosamine residues is also found in some organisms. The peptide side chains always have four amino acid units alternating L-, D-, L-, D- in configuration. The second residue is always D-glutamic acid, linked through its γ-carboxyl group, and the fourth is invariably D-alanine. The peptidoglycan from *Staphylococcus aureus* (type A2) is characteristic of many Gram-positive cocci. Peptidoglycans of this group, and the related types A3 and A4, have similar tetrapeptide side chains but vary in their bridging groups. The amino acids in the bridge are usually glycine, alanine, serine or threonine, and the number of residues can vary from one to five. In type Al peptidoglycans, the L-lysine of the type II peptide side chain is usually replaced by *meso*-2,6,-diaminopimelic acid, and there is no bridging group. Cross-linking occurs between the D-alanine of one side chain and the 6-amino group of the diaminopimelic acid of another. This peptidoglycan type is characteristic of many rod-shaped bacteria, both the large family of Gram-negative rods and the Gram-positive bacilli. In the less common type B peptidoglycans, cross-linkage occurs between the α-carboxyl group of the D-glutamic acid of one peptide side chain and the D-alanine of another through a bridge containing a basic amino acid.

2.3.7 Cross-linking in Gram-negative bacteria

In contrast to the multiple random cross-linkage of peptidoglycan which is found in the Gram-positive cocci, the peptidoglycan of *Escherichia coli* and similar Gram-negative rods has on average only a single cross-link between one peptide side chain and another. These bacteria contain, besides the transpeptidases concerned in cross-linkage, other enzymes known as DD-carboxypeptidases which specifically remove D-alanine from a pentapeptide side chain. Carboxypeptidase I is specific for the terminal D-alanine of the pentapeptide side chain, whilst carboxypeptidase II acts on the D-alanine at position 4 after the terminal D-alanine has been removed. DD-Carboxypeptidase I therefore limits the extent of cross-linking.

The peptidoglycan sacculus determines the overall shape of the cell, and the peptidoglycan is laid down with a definite orientation in which the polysaccharide chains run perpendicular to the main axis of rod-shaped organisms such as *Escherichia coli*.

2.3.8 Attachments to peptidoglycans

Within the cell wall, the polymeric peptidoglycan is usually only part of a larger polymer. In Gram-positive cocci it is linked to an acidic polymer, often a teichoic acid (Figure 2.8). The point of attachment is through

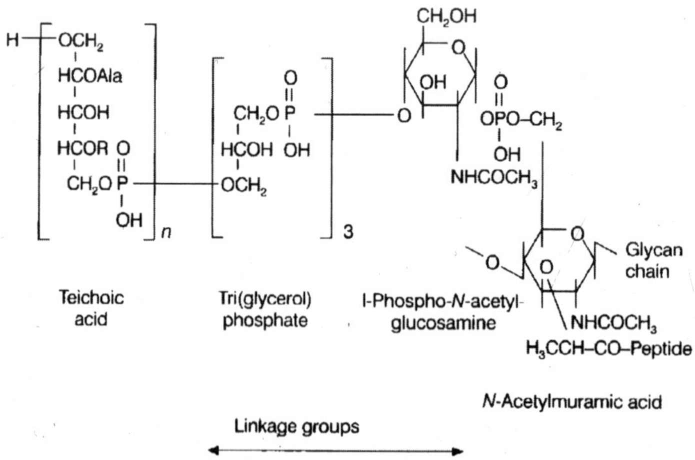

FIGURE 2.8 Teichoic acid and its linkage to peptidoglycan in the wall of *Staphylococcus aureus*.

the 6-hydroxyl group of muramic acid in the glycan chain. Only a small fraction of the muramic acid residues is thus substituted. In *Staphylococcus aureus* cell walls, teichoic acid is joined to peptidoglycan by a linking unit consisting of three glycerol 1-phosphate units attached to the 4 position of *N*-acetylglucosamine which engages through a phosphodiester group at position 1 with the 6-hydroxyl group of muramic acid. This type of linkage seems to occur with polymers other than teichoic acid, e.g. with poly(*N*-acetylglucosamine 1-phosphate) in a *Micrococcus* species. The acid-labile *N*-acetylglucosamine 1-phosphate linkage and the alkali-labile phosphodiester linkage at position 4 explain the ease with which teichoic acid can be split off from peptidoglycan. Within the cell wall, the synthesis of teichoic acid is closely associated with that of peptidoglycan.

In the Gram-positive mycobacteria, the peptidoglycan carries quite a different polymeric attachment. Arabinogalactan is attached to the 6 position of some of the *N*-glycolylmuramic acid residues of the glycan chain through a phosphate ester group. Mycolic acids (complex, very long-chain fatty acids) are attached by ester links to the C-5 position of arabinose residues of the arabinogalactan.The mycobacterial cell wall thus has a high lipid content.

In *Escherichia coli* and related bacteria, the peptidoglycan carries a lipoprotein as a substituent (Figure 2.9). The lipoprotein consists of a polypeptide chain of 58 amino acid units of known sequence with lysine at the C-terminal and cysteine at the N-terminal. This is attached to the 2-carboxyl group of *meso*-2,6-diaminopimelic acid in a peptide side chain of *Escherichia coli* peptidoglycan which has lost both D-alanine groups. Attachment is by an amide link with the ε-amino group in the terminal lysine of the polypeptide. At the opposite end of the polypeptide chain, the cysteine amino group carries a long-chain fatty acid joined as an amide, and its sulfur atom forms a thioether link with a long-chain diacylglycerol.

Lipoprotein occurs in enteric bacteria other than *Escherichia coli*, but it may not be common to all Gram-negative bacteria, although small amounts have been detected in *Proteus mirabilis*.

2.4 Antibiotics that inhibit peptidoglycan biosynthesis

The conclusion that a particular antibiotic owes its antibacterial activity to interference with peptidoglycan biosynthesis rests on several lines of evidence:

1. Bacteria suspended in a medium of high osmotic pressure are protected from concentrations of the antibiotic that would cause lysis and death in a normal medium. Under these conditions the cells lose the shape-determining action of the peptidoglycan and become spherical; they are then known as spheroplasts. These retain an undamaged cytoplasmic membrane, but their wall is deficient or considerably modified. Spheroplasts are in

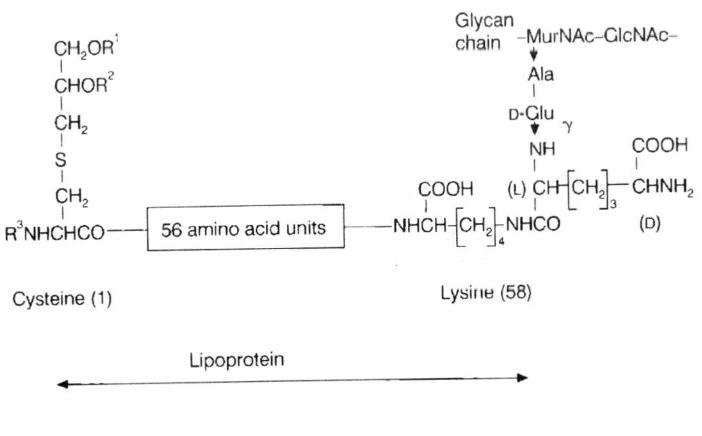

FIGURE 2.9 Lipoprotein and its linkage to peptidoglycan in the envelope of *Escherichia coli*.

principle viable and if the antibiotic is removed, they can divide and produce progeny with normal walls.

2. Several species of bacteria have walls containing no peptidoglycan. These include the mycoplasmas, the halophilic bacteria tolerant of high salt concentrations and bacteria in the L-phase where the normal wall structure is greatly modified. If a compound inhibits the growth of common bacteria but fails to affect bacteria of these special types, it probably owes its activity to interference with peptidoglycan synthesis.
3. Subinhibitory concentrations of these antibiotics often cause accumulation in the bacterial cytoplasm of uridine nucleotides of *N*-acetylmuramic acid, with varying numbers of amino acid residues attached which represent intermediates in the early stages of peptidoglycan biosynthesis. When an antibiotic causes a block at an early point in the reaction sequence, it is not surprising to find an accumulation of the intermediates immediately preceding the block. However, quantities of muramic acid nucleotides are also found in bacteria treated with antibiotics known to affect later stages in peptidoglycan biosynthesis. It seems that all the biosynthetic steps associated with the membrane are closely interlocked, and inhibition of any one of them leads to accumulation of the last water-soluble precursor, UDP-*N*-acetylmuramyl pentapeptide (V, Figure 2.5).

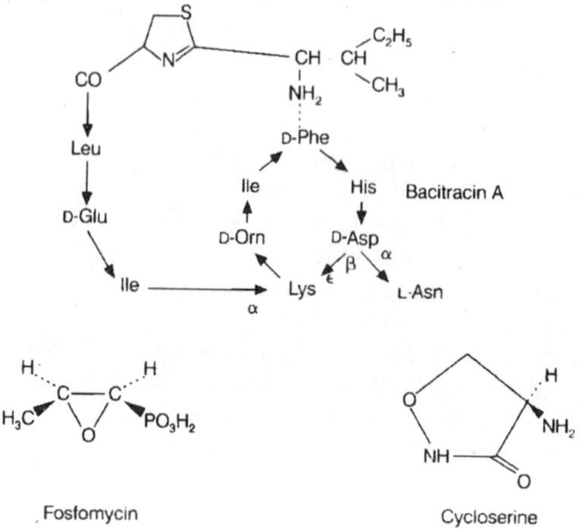

FIGURE 2.10 Antibiotics which inhibit the biosynthesis of the precursors of peptidoglycan.

2.4.1 Bacitracin

Bacitracin is a polypeptide antibiotic (Figure 2.10) which is too toxic for systemic administration but is sometimes used topically to kill Gram-positive bacteria by interfering with cell wall biosynthesis. The antibiotic is ineffective against Gram-negative bacteria, probably because its large molecular size hinders penetration through the outer membrane to its target site. Bacitracin inhibits peptidoglycan biosynthesis by binding specifically to the long-chain C_{55}-isoprenol pyrophosphate in the presence of divalent metal ions.

In the formation of the linear peptidoglycan (IX, Figure 2.7), the membrane-bound isoprenol pyrophosphate is released. Normally this is converted by a pyrophosphatase to the corresponding phosphate which thus becomes available for reaction with another molecule of UDPMur-*N*-Ac-pentapeptide (V, Figure 2.6). Interaction between the lipid pyrophosphate and a metal ion-bacitracin coordination complex blocks this process and eventually halts the synthesis of peptidoglycan. The identity of the divalent metal ion bound to the antibiotic in bacterial cells is uncertain but could well be either Mg^{2+} or Zn^{2+}. Bacitracin forms 1:1 complexes with several divalent metal ions, and investigations employing nuclear magnetic resonance and optical rotary dispersion (ORD) indicate the involvement of the imidazole ring of the histidine residue of the antibiotic in metal ion binding. Additional likely sites of metal ion interaction include the thiazoline moiety and the carboxyl groups of the D-aspartate and D-glutamate residue.

2.4.2 Fosfomycin (phosphonomycin)

This antibiotic has the very simple structure shown in Figure 2.10. It acts on infections caused by both

Gram-positive and Gram-negative bacteria but although its toxicity is low, until recently it achieved only limited use in clinical practice. However, there is a resurgence of interest in fosfomycin for the treatment of serious infections resistant to other antibiotics. Fosfomycin inhibits the first step of peptidoglycan biosynthesis, namely, the condensation of UDP-*N*-acetylglucosamine (I) with phosphoenol pyruvate (PEP) catalyzed by UDP-*N*-acetylglucosamine enolpyruvyl transferase (MurA), giving the intermediate (II) that subsequently yields UDP-*N*-acetylmuramic acid (III) on reduction (Figure 2.4). Fosfomycin inactivates MurA by reacting covalently with an essential cysteine residue (Cys-115) at the active center of the enzyme to form the thioester illustrated in Figure 2.11. This reaction is time-dependent and is facilitated by UDP-*N*-acetylglucosamine, which appears to 'chase' the other substrate (PEP) from the active site and promotes a conformational change in the enzyme. Both these effects are believed to expose the nucleophilic Cys-115 for reaction with the epoxide moiety of fosfomycin. The three-dimensional structure of MurA (from *Escherichia coli*) complexed with UDP-*N*-acetylglucosamine and fosfomycin has been determined by X-ray crystallography. The analysis confirmed the covalent interaction of the antibiotic with Cys-115 and also revealed that there are hydrogen bonds between the antibiotic and the enzyme and UDP-*N*-acetylglucosamine.

2.4.3 Cycloserine

This antibiotic also has a simple structure (Figure 2.10). Cycloserine is active against several bacterial species, but because of the central nervous system disturbances which are experienced by some patients, clinically it is limited to occasional use in individuals with tuberculosis that is resistant to the more commonly used drugs. Cycloserine produces effects in bacteria that are typical of compounds acting on peptidoglycan biosynthesis. Thus when cultures of *Staphylococcus aureus* are grown with subinhibitory concentrations of cycloserine, the peptidoglycan precursor (IV, Figure 2.5) accumulates in the medium, suggesting a blockage in the biosynthesis immediately beyond this point.

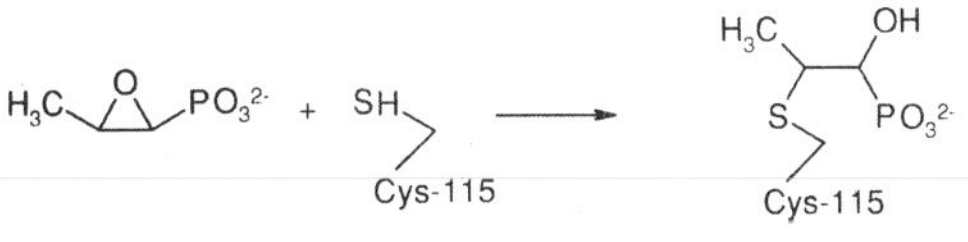

FIGURE 2.11 Fosfomycin inactivates UDP-*N*-acetylglucosamine enolpyruvoyl transferase (MurA) by reacting with the essential cysteine residue (Cys-115) at the active center of the enzyme to form a thioester.

In fact, cycloserine inhibits alanine racemase and D-alanyl-D-alanine ligase, the two enzymes concerned in making the dipeptide for completion of the pentapeptide side chain. Molecular models reveal that cycloserine is structurally related to one possible conformation of D-alanine, so that its inhibitory action on these enzymes appears to be a classic example of isosteric interference. The observation that the action of cycloserine is specifically antagonized by the addition of D-alanine to the growth medium also supports the postulated site of action. The affinity of cycloserine for the ligase is much greater than that of the natural substrate, the ratio of K_m to K_i being about 100. In a compound acting purely as a competitive enzyme inhibitor, this sort of K_m/K_i ratio is probably essential for useful antibacterial activity. The greater affinity of cycloserine for the enzyme may be connected with its rigid structure. This could permit a particularly accurate fit to the active center of the enzyme, either in the state existing when the enzyme is uncombined with its substrate or in a modified conformation which is assumed during the normal enzymic reaction. Rigid structures of narrow molecular specificity are common among antimicrobial agents and similar considerations may apply to other types of action; this theme will recur in later sections. The three-dimensional structures of both alanine racemase and D-alanyl-D-alanine ligase are available and it will be interesting to see whether cycloserine does indeed interact with the active sites of these enzymes according to this concept of inhibition.

Cycloserine enters the bacterial cell by active transport (see Chapter 7). This allows the antibiotic to reach higher concentrations in the cell than in the medium and adds considerably to its antibacterial efficacy.

2.4.4 Glycopeptide antibiotics

Vancomycin (Figure 2.12), which is a member of a group of complex glycopeptide antibiotics, was first isolated in the 1950s, but its real clinical importance only emerged with the inexorable spread of methicillin-resistant staphylococci (MRSA; see Chapters 9 and 10). The use of vancomycin and structurally related glycopeptides has markedly increased because of their value in treating serious infections caused by MRSA and other Gram-positive bacteria. Because of their relatively large molecular size, the glycopeptides are essentially inactive against the more impermeable Gram-negative bacteria. The antibacterial action of glycopeptide antibiotics depends on their ability to bind specifically to the terminal D-alanyl-D-alanine group on the peptide side chain of the membrane-bound intermediates in peptidoglycan synthesis (compounds VI–IX in Figure 2.6). It is important to note that this interaction occurs on the outer face of the cytoplasmic membrane. The glycopeptide antibiotics probably do not enter the bacterial cytoplasm, again because of their molecular size. The complex which is formed between vancomycin and D-alanyl-D-alanine

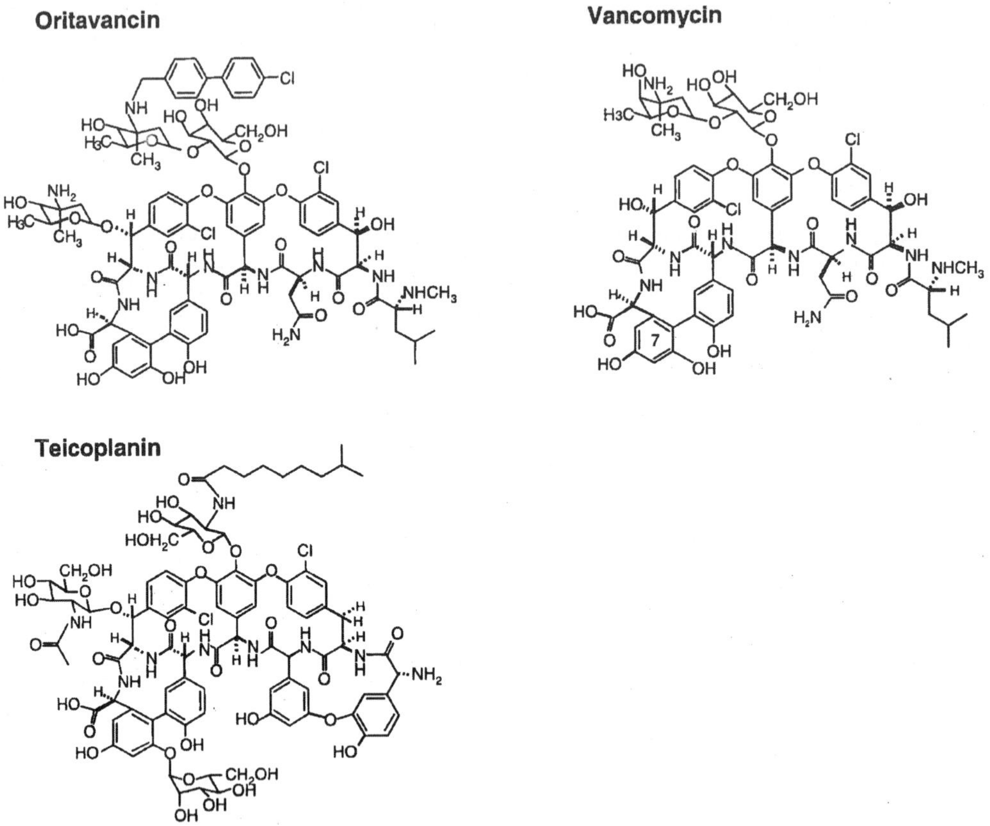

FIGURE 2.12 Glycopeptide antibiotic inhibitors of peptidoglycan synthesis that are increasingly important in the treatment of infections caused by drug-resistant staphylococci.

has been studied in considerable detail. The complex blocks the transglycosylase involved in the incorporation of the disaccharide-peptide into the growing peptidoglycan chain and the DD-transpeptidases and DD-carboxypeptidases for which the D-alanyl-D-alanine moiety is a substrate. Both peptidoglycan chain extension and cross-linking are therefore inhibited by glycopeptide antibiotics. This is, in fact, a most unusual mode of inhibition in that the antibiotic prevents the utilization of the substrate rather than directly interacting with the target enzymes.

The side chains of the amino acids of the heptapeptide backbone of vancomycin are extensively cross-linked to form a relatively concave carboxylate cleft into which the D-alanyl-D-alanine entity binds noncovalently via hydrogen bonds and hydrophobic interactions. Furthermore, NMR and X-ray crystallographic studies show that vancomycin spontaneously forms a dimeric structure which enables the antibiotic to bind to two D-alanyl-D-alanine peptide units attached either to the disaccharide-peptide precursor or to adjacent growing peptidoglycan strands. Another glycopeptide antibiotic, teichoplanin (Fig 2.12), is considerably more potent than vancomycin against some important Gram-positive pathogens. It is thought that the *N*-substituted fatty acyl side chain that distinguishes teichoplanin from vancomycin serves to anchor teichoplanin in the cytoplasmic membrane. This localization may facilitate the interaction of the drug with the D-alanyl-D-alanine target site. In contrast with vancomycin, teichoplanin does not form dimers. Thus although the dimerization of vancomycin may in principle facilitate its antibacterial action, the dimerizing potential is relatively weak and it is unclear whether the dimer is indeed a significant contributor to the antibiotic activity of vancomycin *in vivo*. The semisynthetic glycopeptide, oritavancin (Fig 2.12), is strongly dimerized and this may be a factor in the highly potent antibacterial activity of this promising drug.

2.4.5 Penicillins, cephalosporins and other β-lactam antibiotics

Penicillin was the first naturally occurring antibiotic to be used for the treatment of bacterial infections, and the story of its discovery and development is one of the most inspiring in the history of medicine. Penicillin is one of a group of compounds known as β-lactam antibiotics which are unrivalled in the treatment of bacterial infections. Their only serious defects include an ability to cause immunologic sensitization in a small proportion of patients, a side effect which prevents their use in those affected, and the frequency of emergence of bacteria resistant to β-lactams. The original penicillins isolated directly from mold fermentations were mixtures of compounds having different side chains. The addition of phenylacetic acid to the fermentation medium improved the yield of penicillin and ensured that the product was substantially a single compound known as penicillin G or benzylpenicillin (Figure 2.13). The first successful variant was obtained by replacing phenylacetic acid by phenoxyacetic acid as the added precursor. This gave phenoxymethylpenicillin or penicillin V (Figure 2.13). The main advantage of this change was an improvement in the stability of the penicillin towards acid. The ready inactivation of penicillin G at low pH limited its usefulness when it was given by mouth since a variable and often considerable fraction of the antibacterial activity was destroyed in the acidic environment of the stomach. Penicillin V thus improved the reliability of oral doses. These early penicillins, produced directly by fermentation, were intensely active against Gram-positive infections and gave excellent results in streptococcal and staphylococcal infections and in pneumonia. They were also very active against Gram-negative infections caused by gonococci and meningococci, but were much less active against the more typical Gram-negative bacilli.

A further advance in the versatility of the penicillins was achieved by workers at the original Beecham company (now part of GlaxoSmithKline) with the development of a method for the chemical modification of the penicillin molecule. Bacterial enzymes were found that remove the benzyl side chain from penicillin G, leaving 6-aminopenicillanic acid, which could be isolated and then acylated by chemical means. This discovery opened the way to the production of an almost unlimited number of penicillin derivatives, some of which have shown important changes in properties compared with the parent penicillin. The value of increased stability has already been mentioned, and some semisynthetic penicillins show this

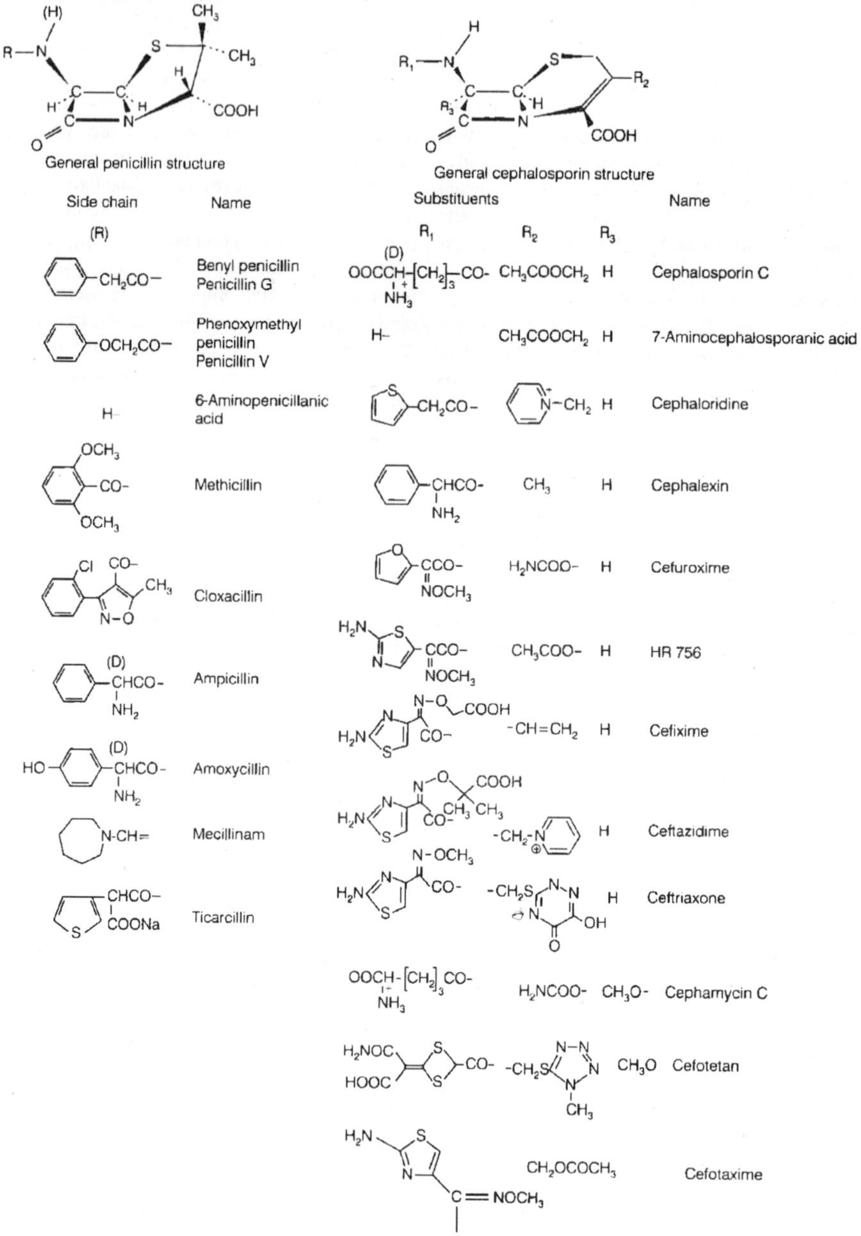

FIGURE 2.13 Representative penicillins and cephalosporins.

property. Other modified penicillins (e.g. methicillin and cloxacillin, Figure 2.13) are much less susceptible to attack by β-lactamase, an enzyme which converts penicillin to the antibacterially inactive penicilloic acid and gives rise to the commonest form of resistance to penicillin (Chapter 9).

The discovery of the β-lactamase inhibitor, clavulanic acid (Figure 2.14), which is a β-lactam itself but without useful antibacterial activity, provided an opportunity to coadminister this agent with β-lactamase-sensitive compounds such as amoxycillin (Figure 2.13) in mixtures such as augmentin (a 1:1 mixture of amoxycillin and clavulanic acid) and timentin (a 1:1 mixture of ticarcillin and clavulanic acid).

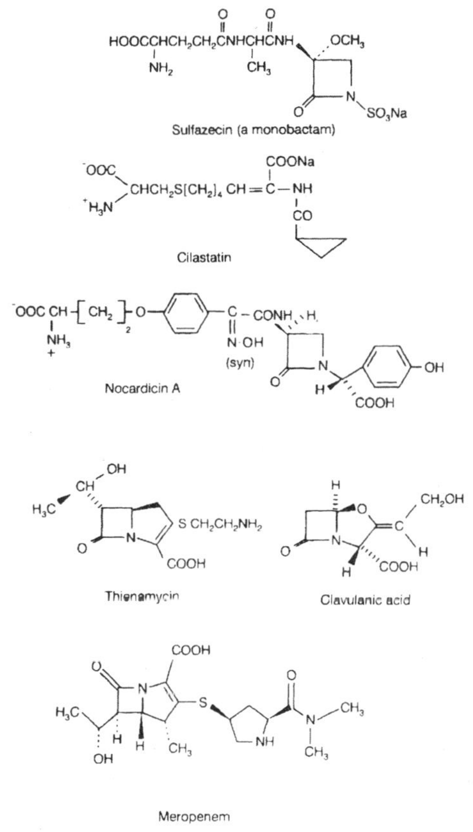

FIGURE 2.14 Additional β-lactam compounds and cilastatin, an inhibitor of mammalian metabolism of thienamycin. Clavulanic acid is an inhibitor of serine-active-site β-lactamases.

Another striking change brought about by chemical modification of the penicillin side chain was an increase in activity against Gram-negative bacteria, a property found in several derivatives, including ampicillin, amoxycillin, carbenicillin and ticarcillin (Figure 2.13). This increase in Gram-negative activity is accompanied by a lessening of activity towards Gram-positive bacteria. Ampicillin is one of the most widely used antibacterial agents. In mecillinam (Figure 2.13), where the side chain is attached by an azomethine link rather than the usual amide bond, the activity spectrum of the original penicillin molecule has been completely reversed. This compound is highly active against Gram-negative bacteria but requires 50 times the concentration for an equal effect on Gram-positive organisms. It can be used in the treatment of typhoid fever, which is caused by the Gram-negative bacterium *Salmonella typhi*.

Cephalosporin C (Figure 2.13), originally isolated from a different organism than that used to produce penicillin, has a structure in its nucleus similar to that in the penicillins. The biogenesis of the nuclei in these two classes of antibiotics is now known to be identical except that in cephalosporin biosynthesis the thiazolidine ring of the penicillin nucleus undergoes a specific ring expansion to form the dihydrothiazine ring of the cephalosporin nucleus. Besides this similarity in structure and biogenesis, cephalosporin C and its derivatives act on peptidoglycan cross-linking in the same way as the penicillins. Cephalosporin C itself is not a useful antibacterial drug, but like the penicillins, it is amenable to chemical modification. Enzymic removal of the side chain gives 7-aminocephalosporanic acid, which can be chemically acylated to give new derivatives. A second change in the molecule can also be made by a chemical modification of the acetoxy group of cephalosporin C. The first successful semisynthetic cephalosporin was cephaloridine. Many others have followed; a selection of some the best known is shown in Figure 2.13. Most are only effective when given by injection, but cephalexin and cefixime can be given by mouth. Cefuroxime is unaffected by many of the common β-lactamases and can be used against bacterial strains which are resistant to other β-lactam antibiotics; it can be useful in infections that are due to *Neisseria* or *Haemophilus*. The related compound, cefotaxime, has enjoyed considerable success. Other agents such as ceftazidime and ceftriaxone are useful because of the former's improved antipseudomonal activity and the latter's enhanced half-life in the body, which permits a more convenient dosing schedule, for example, once or twice daily.

The cephamycins resemble the cephalosporins, but have a methoxy group in place of hydrogen at position 7. Cefotetan (Figure 2.13) is a semisynthetic derivative of cephamycin C. The cephamycin derivatives are not readily attacked by β-lactamases and have advantages over the cephalosporin derivatives, with activity against *Proteus* and *Serratia* species.

The enormous success of the penicillins and cephalosporins stimulated a search for other naturally occurring β-lactam compounds. These have been found in a variety of micro-organisms. Some of the most interesting are shown in Figure 2.14. In the carbapenem, thienamycin, the sulfur atom is not part of the ring, but is found in the side chain. This compound is remarkable for its high potency, broad antibacterial spectrum and resistance to β-lactamase attack, but it is both chemically unstable and susceptible to degradation by a dehydropeptidase found in the kidneys. The *N*-formimidoyl derivative of thienamycin is chemically more stable but must administered as a 1:1 mixture with cilastatin (Figure 2.14), an inhibitor of the renal peptidase. A further development in the carbapenem series has been the appearance of the synthetic compound meropenem (Figure 2.14). This drug is not readily degraded by renal peptidase and can therefore be administered as a single agent. Meropenem is active against Gram-positive and Gram-negative pathogens, including many which are resistant to other β-lactams.

Other β-lactam antibiotics include the monobactams (e.g. sulfazecin, Figure 2.14); the name comes from monocyclic bacterial β-lactams) which are derived from bacteria and represent the simplest β-lactam structures with antibacterial activity so far discovered. Many semisynthetic derivatives have been made and exhibit excellent anti-Gram-negative activity, with much weaker activity against Gram-positive bacteria. In contrast, the monocyclic nocardicins (Figure 2.14) appear to offer less activity and are of more historic than clinical interest. Interest in the β-lactam family remains intense and novel drugs with improved properties continue to be developed.

2.4.6 Mode of action of penicillins and cephalosporins

As with many other antibiotics, early attempts to discover the biochemical action of penicillin led to conflicting hypotheses. Gradually it became accepted that the primary site of action lay in the production of cell wall material, and more specifically in the biosynthesis of peptidoglycan.

Evidence for this site of action rests on several different types of experiment. *Staphylococcus aureus* cells were pulse-labelled with [^{14}C]glycine, and peptidoglycan was isolated from their walls after a further period of growth in unlabelled medium. The labelled glycine entered the pentaglycyl 'extending group'. The polysaccharide backbone of the peptidoglycan was then broken down by an *N*-acetylmuramidase, leaving the individual muramyl peptide units linked only by their pentaglycine peptide chains. After the products were separated by gel chromatography, radioactivity was found in a series of peaks of increasing molecular weight representing the distribution of the pulse of [^{14}C]glycine among peptide-linked oligomers of varying size. A parallel experiment done in the presence of penicillin showed the radioactivity to be associated largely with a single peak of low molecular weight, presumably the un-cross-linked muramyl peptide unit, with much less radiolabel in the oligomers. The penicillin had thus inhibited the peptide cross-linking.

In another experiment, 'nucleotide pentapeptide' was prepared with [^{14}C]alanine. This was used as a substrate for an enzyme preparation from *Escherichia coli* in the presence of UDP-*N*-acetylglucosamine. This system carried out the entire biosynthesis of peptidoglycan, including the final stage of cross-linking. Peptidoglycan was obtained as an insoluble product containing [^{14}C] from the penultimate D-alanine of the substrate; the terminal D-[^{14}C]alanine was released into the medium, partly from the transpeptidase cross-linking reaction and partly from a carboxypeptidase that removed terminal D-alanine residues from cross-linked products. In a parallel experiment, penicillin was added at a concentration that would inhibit growth of *Escherichia coli*. Biosynthesis of peptidoglycan then proceeded only to the stage of the linear polymer (IX, Figure 2.6), which was isolated as a water-soluble product of high molecular weight labelled with [^{14}C]. No D-[^{14}C]alanine was liberated because the penicillin suppressed both the cross-linking transpeptidase reaction and the action of DD-carboxypeptidase.

The understanding of the mechanism of β-lactam action was considerably advanced by the discovery of the penicillin-binding proteins referred to in Section 2.3.5. Of the PBPs in *Escherichia coli* and many other bacteria, PBP1a and PBP1b are the key enzymes involved in peptidoglycan biosynthesis. PBP2 and PBP3 are concerned, respectively, with remodelling of the peptidoglycan sacculus during septation and cell division. All these PBPs are targets of β-lactam antibiotics. Different β-lactams exhibit different affinities for the various PBPs and these can in turn be correlated with different morphological effects. Drugs which bind most strongly to PBPs 1a and 1b cause cell lysis at the lowest antibacterial concentration. Compounds such as the cephalosporin, cephalexin, bind more strongly to PBP3 and inhibit septation, leading to the formation of filaments, which are greatly elongated cells. Another variation is found with mecillinam, which binds almost exclusively to PBP2 and causes cells to assume an abnormal ovoid shape. Cells overproducing PBP2 have enhanced amounts of cross-linked peptidoglycan and are very sensitive to mecillinam.

The interaction of a penicillin or cephalosporin (I) with the enzyme (E) can be represented as:

$$\mathrm{E} + \mathrm{I} \underset{k_2}{\overset{k_1}{\rightleftarrows}} \mathrm{EI} \xrightarrow{k_3} \mathrm{EI}^* \xrightarrow{k_4} \mathrm{E} + \text{degraded inhibitor.}$$

The first step is reversible binding to the enzyme. The second stage, involving chemical modification of the inhibitor with covalent binding to the enzyme, is irreversible, as is the final stage of enzyme release. For high antibacterial activity, k_3 should be rapid, preventing release of inhibitor through reversal of the initial binding, and k_4 should be slow to maintain the enzyme in the inactive EI* form and to avoid significant reactivation. Measurements show that the widely used β-lactam antibiotics have just such characteristics, and this scheme goes far to explain their outstanding effectiveness. There is good reason to suppose that the inactivation mechanism is the same with cross-linking transpeptidases as with DD-carboxypeptidases. The nature of the end products of penicillin degradation depends on the enzyme involved. It may be a simple opening of the β-lactam ring to give the penicilloate or there may be more extensive breakdown leading to the production, from benzylpenicillin, of phenylacetyl glycine. Those enzymes which yield penicilloate are equivalent to slow-acting β-lactamases. There is evidence to suggest that active β-lactamases are relatives of carboxypeptidases and transpeptidases in which reaction k_4 is rapid instead of very slow.

The mechanism of action of DD-carboxypeptidases and cross-linking transpeptidases resembles that of certain esterases and amidases. These enzymes possess reactive groups associated with their active centres, which undergo transient acylation in the course of enzymic action. Antibiotics containing a β-lactam ring behave chemically as acylating agents. The action of penicillin on the PBPs thus involves acylation of the enzymically active site in the second reaction to form the inactive complex EI*. This explanation was supported by experiments with purified DD-carboxypeptidases from *Bacillus stearothermophilus* and *Bacillus subtilis*. The enzyme was allowed to react briefly with [^{14}C]benzylpenicillin or with a substrate analogue, [^{14}C]Ac$_2$L-Lys-D-Ala-D-lactate; D-lactic acid is the exact hydroxyl analogue of D-alanine, and use of this derivative enabled the transient enzyme reaction intermediate to be trapped. In peptide fragments from the *Bacillus stearothermophilus* enzyme, radioactivity was found in a peptide with 40 amino acid residues and the label was shown to be associated with the same specific serine residue, whether the reactant was benzylpenicillin or the substrate analogue. Similar results were found with the *Bacillus subtilis* enzyme from which a labelled 14-unit peptide was isolated. This peptide showed extensive homology with 14 residues of the *Bacillus stearothermophilus* peptide and the label was associated with the corresponding serine residue. It was concluded that penicillin binds to the active site and acylates the same serine as the substrate. Unlike the substrate, the degraded penicillin was released very slowly (reaction k_4 in the scheme shown) and thus blocked further access of substrate to the site.

How can this action of penicillin be related to its structure? The most widely quoted explanation depends on the similarity of the spatial orientation of the principal atoms and polar groups in the β-lactam nucleus to one particular orientation of the D-alanyl-D-alanine end group of the pentapeptide side chain of peptidoglycan precursors (see Figure 2.15). When the two structures are compared, the peptide bond between the alanine units is seen to correspond in position to the C—N bond in the β-lactam ring which is believed to be responsible for the acylating activity. Such a group bound to the cross-linking transpeptidase close to its active centre could well usurp the acylating function implicit in the normal reaction of the substrate with the enzyme. When the structures (Figure 2.15) are compared more critically, it becomes apparent that the agreement between them is imperfect but can be much improved if the peptide bond of the D-alanyl-D-alanine end group is represented, not in its normal planar form, but twisted nearly 45° out of plane. This may imply that the conformation of the penicillin molecule resembles the transition state of the substrate rather than its resting form. During the enzymic transpeptidation, the peptide bond quite possibly undergoes this sort of distortion.

An alternative model is based on a comparison of certain electrostatic potentials of benzylpenicillin and synthetic *N*-acyl-D-alanyl-D-alanine peptides. Calculation of these potentials reveals a significant similarity in the coplanarity of key electrostatic negative wells of both benzylpenicillin and the dipeptide terminal. The coplanarity of these wells may facilitate the attack of an electrophilic centre in the catalytically active serine of the target PBP on the β-lactam C—N bond. With some modifications this model may be applicable to all

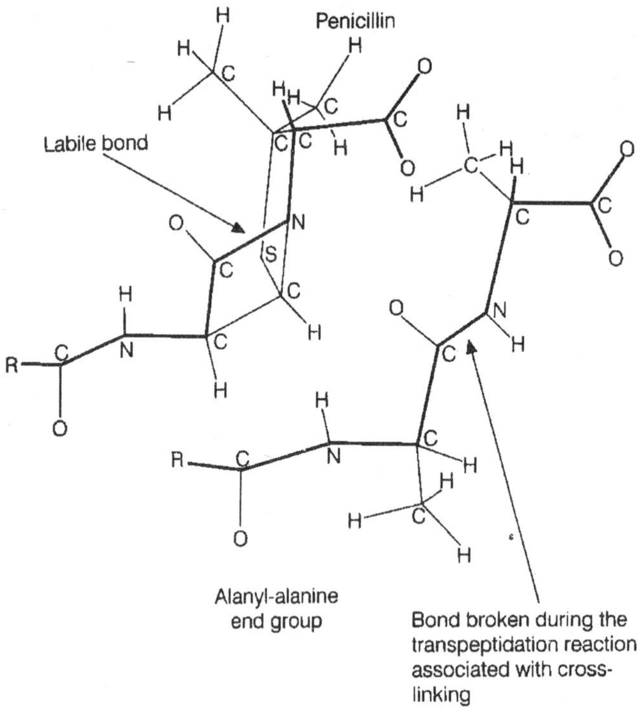

FIGURE 2.15 Comparison of the structures of penicillin with that of the D-alanyl-D-alanine end group of the peptidoglycan precursor. [Reproduced by permission of the Federation of American Societies for Experimental Biology from J. L. Strominger *et al. Fed. Proc.* **26**, 18 (1967).]

types of β-lactam drug. However, the precise details of the interactions between β-lactam and PBPs will have to await data from the various on-going X-ray crystallographic studies of β-lactam–PBP complexes.

2.5 Drugs that interfere with the biosynthesis of the cell wall of mycobacteria

Mycobacteria are responsible for two devastating diseases: tuberculosis (*Mycobacterium tuberculosis*) and leprosy, or Hansen's disease (*Mycobacterium leprae*). The cell wall of mycobacteria is remarkably complex and underlies many of the characteristic properties of these organisms, including their extremely low permeability and intrinsic resistance to commonly used antibiotics. The reader is referred to a review provided in 'Further reading' at the end of this chapter for detailed information on the cell wall of mycobacteria. A key feature of the mycobacterial cell envelope that distinguishes it from most other bacteria is the mycolyl–arabinogalactan–peptidoglycan complex. Arabinogalactan is linked to the peptidoglycan through a phosphodiester link between the C-6 of 10–14% of the muramic acid residues and a disaccharide linker unit attached to the galactan. Arabinogalactan itself is a unique polysaccharide consisting of linear galactan chains composed of alternating 5- and 6-linked β-D-galactofuranose units which in turn are linked though C-5 of some of the 6-linked galactofuranose units to extensively branched chains of D-arabinofuranose (arabinan). Approximately two-thirds of the nonreducing terminals of arabinan are esterified to long-chain mycolic acids. There are other lipids in the mycobacterial cell outer envelope in addition to the mycolic acids, including a range of complex glyco- and peptidolipids. The lipoidal nature of this complex wall is a significant contributor to the impermeability of mycobacteria to many solutes, including some antibiotics. The characteristically slow growth rate of mycobacteria also presents a considerable challenge to the successful chemotherapy of infections caused by these bacteria, which usually requires several months of continuous drug treatment.

2.5.1 Isoniazid

Isoniazid (Figure 2.16) provides one of the foundations of combination therapy for tuberculosis. The relative ease with which *Mycobacterium tuberculosis* becomes resistant to individual drugs led to the concept of combining several chemically distinct drugs with, as it later turned out, different modes of action. In combination variously with rifampicin, ethambutol, pyrazinamide and occasionally streptomycin, isoniazid is an effective antitubercular drug which has been in use since 1952. However, it is only since the 1990s that biochemical and genetic data have revealed the molecular mechanisms underlying the antimycobacterial action of isoniazid.

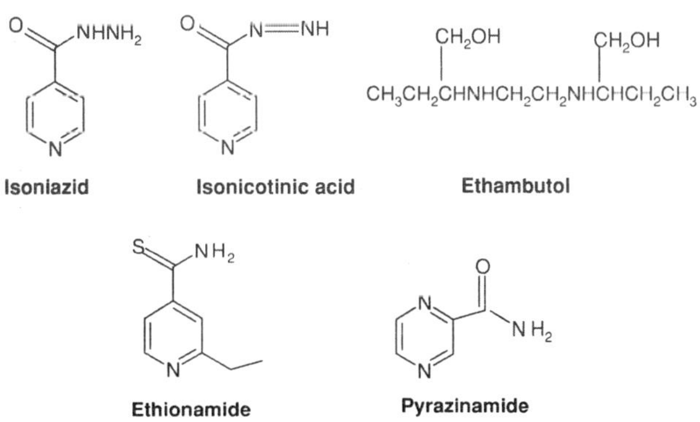

FIGURE 2.16 Structures of synthetic compounds used in combination therapy of tuberculosis. The structure of the microbial metabolite isonicotinic acid can be seen to resemble that of isoniazid.

In the mycobacteria, the *inhA* gene encodes an enzyme that has been identified as a major molecular target for isoniazid and the structurally related drug, ethionamide (Figure 2.16). This enzyme, abbreviated to InhA, catalyzes the NADH-dependent reduction of the 2-*trans*-enoyl-acyl carrier protein (ACP), an essential reaction in the elongation of fatty acids. Long-chain substrates containing between 16 and 18 carbon atoms are preferentially used by InhA, an observation which implicates the reductase in the biosynthesis of the mycolic acids. Inhibition of the biosynthesis of mycolic acids therefore disrupts the assembly of the mycolyl-arabinogalactan-peptidoglycan complex and causes the loss of cell viability. While mutations in *inhA* confer resistance to isoniazid, studies with recombinant InhA show that isoniazid itself is only a weak inhibitor of the enzyme. The drug is in fact first converted by oxidative cellular metabolism to a reactive metabolite which is believed to bind to and inhibit the reductase in the presence of NADH bound to the enzyme. Isoniazid is metabolically unstable in mycobacteria, owing to the activity of a unique mycobacterial catalase–peroxidase encoded by the *katG* gene. Studies with the recombinant form of this enzyme show that it converts isoniazid to several chemically reactive derivatives, isonicotinic acid (Fig 2.16) being the major product. The electrophilic nature of these compounds would enable them to acylate or oxidize vulnerable amino acid residues in the target reductase, although direct evidence for this is lacking. The two-stage concept of the mechanism of action of isoniazid is strengthened by the existence of two forms of mycobacterial resistance to the drug. One type of mutant has a defective *katG* gene that precludes the conversion of the prodrug to its active form. The second resistant phenotype depends on an isoniazid-resistant variant of InhA that is characterized by a markedly lower affinity for NADH, which minimizes the attack of the isoniazid metabolite on the enzyme.

Recently, the product of another gene, *kasA*, has been proposed as an alternative primary site of action for isoniazid. The *kasA* gene encodes the enzyme β-ketoacyl synthase (KasA), which may be involved in the biosynthesis of C_{18}–C_{34} fatty acids required for the elaboration of mycolic acids. The case for the KasA enzyme as a primary target for isonaizid rests largely on isoniazid-resistant clinical isolates of mycobacteria with mutations solely in the *kasA* gene, i.e. with no mutations in either *inhA* or *katG* genes. However, there is contrary evidence of clinical isolates with mutations in *kasA* which retain sensitivity to isoniazid. In summary, the weight of experimental and observational evidence supports the concept of the InhA enzyme as the primary target for isoniazid (and also ethionamide), although a possible contribution from KasA cannot be ruled out at this stage.

2.5.2 Ethambutol

The antibacterial activity of isoniazid is confined to *Mycobacterium tuberculosis*. Ethambutol (Figure 2.16), which has been in clinical use against tuberculosis since 1961, has a broader spectrum of action, including *Mycobacterium avium*, a serious opportunist pathogen in patients with AIDS. Despite many years of use, the molecular basis of the bacteriostatic action of ethambutol was identified only recently. It had long been known that the drug in some way blocked the biosynthesis of the polysaccharide arabinan, but the actual mechanism was not known. The target for ethambutol was eventually established by cloning the genetic elements responsible for resistance to this drug in *Mycobacterium avium*. The structural genes *embA*, *embB* and *embC* all encode arabinosyl transferases which appear to have similar functions in polymerizing arabinose into arabinan. *In vitro* evidence obtained with a crude broken cell preparation from *Mycobacterium smegmatis* indicates that ethambutol inhibits the transfer of a hexa-arabinosfuranosyl unit, from the phospho-decaprenol carrier complex, to arabinan. Most clinical isolates of *Mycobacterium tuberculosis* that are resistant to ethambutol have mutations in *embB*. The molecular target of ethambutol therefore seems to be arabinosyl transferase, with the product of the *embB* gene being the most important. The mechanism of inhibition of the arabinosyl transferases by ethambutol remains to be established and will probably await the purification of the enzymes, which are predicted to be integral membrane proteins with multiple anchoring, transmembrane domains and an external domain.

Disruption of the biosynthesis of the arabinogalactan component of the mycobacterial cell envelope may increase cellular permeability to other drugs. This could account for the valuable clinical synergism that is achieved when ethambutol is combined with a drug of large molecular size such as rifampicin.

2.5.3 Pyrazinamide

Although first recognized for its substantial antimycobacterial activity in the 1950s, this synthetic drug was not introduced into the combination therapy for tuberculosis until the mid-1980s. Pyrazinamide (Figure 2.16) is a bacteriostatic agent which is especially useful against semidormant populations of *Mycobacterium tuberculosis* located in acidic intracellular compartments such as the phagolysosomes of macrophages.

The active form of pyrazinamide is believed to be pyrazinoic acid (Figure 2.17), formed by the action of an intracellular bacterial amidase, referred to as pyrazinamidase. Some pyrazinamide-resistant strains of *M. tuberculosis* lack pyrazinamidase activity. Genetic and biochemical evidence strongly suggests that the antibacterial activity of pyrazinamide rests upon the inhibition by pyrazinoic acid of a multifunctional fatty acid synthase (type I) encoded by the *fasI* gene, which results in suppression of mycolic acid biosynthesis. By inhibiting the type I fatty acid synthase (FAS I), pyrizinamide blocks the provision of fatty acid precursors for another fatty acid synthase (FAS II) which has an essential role in the elongation of mycolic acids.

2.6 The fungal cell wall as a target for antifungal drugs

Fungal infections (mycoses) pose an ever-increasing threat to health across the world. Immunocompromised individuals, including AIDS patients, those on immunosuppressive drugs following organ transplantation, cancer patients undergoing chemotherapy, people recovering from major surgery and patients receiving prolonged antibacterial treatment, are all at risk from infections caused by a variety of fungal pathogens. Compared with the wealth of drugs available to treat bacterial infections, the current therapeutic options for fungal infections are much more limited.

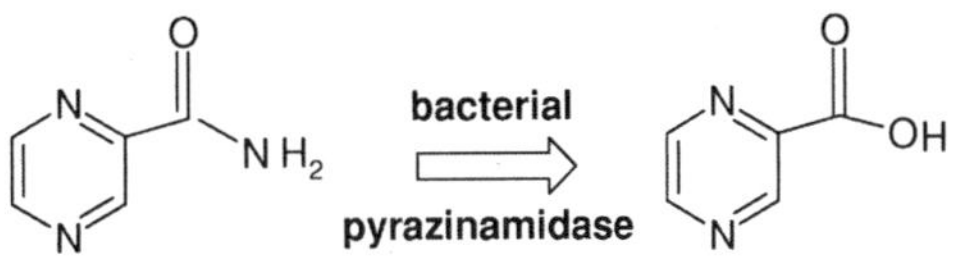

FIGURE 2.17 Conversion of the prodrug pyrazinamide to the active molecule by bacterial amidase.

Although it serves functions analogous to those of the bacterial cell wall, the structure of the fungal wall is very different from that of its bacterial counterpart. Critically, fungal walls do not contain peptidoglycan, so neither β-lactam nor glycopeptide antibiotics have any effect on the viability of fungi. The fungal wall is a multilayered structure whose major macromolecular components include chitin, glucan and mannoproteins. Neither chitin nor glucan occurs in mammalian or bacterial cell walls, so the biosynthesis of these materials provides potential targets for specific antifungal drug action. Because glycosylated proteins are found in all eukaryotes, the biosynthesis of fungal mannoproteins may be rather less attractive as a target for chemotherapy. However, the sugar residues of glycosylated proteins are very different in fungi and humans and could conceivably offer opportunities for drug design. The composition and organization of the cell wall vary significantly among the various fungal species and define the identity of the organisms. Chitin is a linear 1,4-β-linked homopolymer of *N*-acetylglucosamine. In yeasts chitin contributes as little as 2% to the cell wall mass, while there can be as much as 60% chitin in some mycelial fungi. Nevertheless, chitin is essential for fungal growth, even in species with very small amounts of the polymer. Glucan is a β-1,3-linked linear glucose homopolymer with varying amounts of β-1,6- and β-1,4-glucose side chains, depending on the species. The mannoproteins make up complex chains of mannose linearly bonded by 1,6-links to which oligomannoside side branches are attached by 1,2- and 1,2-α bonds. The polysaccharide structures are covalently linked to protein via a 1,4-β-disaccharide of *N*-acetylglucosamine residues by either *N*-glycosylation of asparagine or *O*-glycosylation at the free hydroxyl groups of threonine or serine residues. Some idea of the diversity of mannoproteins may be gauged from the fact that between 40 and 60 different mannoproteins can be isolated from yeast cell walls.

The arrangement of these various polymers in the wall of the important fungal pathogen *Candida albicans* is illustrated in Figure 2.18. The insoluble polymers chitin and glucan confer mechanical strength on the wall. The function of the mannoproteins is less clear but appears to be essential because inhibitors of *N*- and *O*-glycosylation are lethal, although it should be remembered that the effects of inhibition of these reactions are not confined to the biosynthesis of mannoproteins.

2.6.1 Inhibitors of chitin biosynthesis

The enzyme chitin synthase catalyzes a reaction in which an *N*-acetylglucosamine residue is transferred from the donor molecule, UDP-*N*-acetyl glucosamine, to the nonreducing end of the growing chitin chain, with the concomitant release of UDP. Chitin synthase exists in several forms, none of which has been purified so far. Two related groups of antibiotics inhibit chitin synthase, the polyoxins and nikkomycins (Figure 2.19). Both types are analogues of UDP-*N*-acetylglucosamine and presumably inhibit the enzyme by competition with this substrate. In the yeast *Saccharomyces cerevisae,* the gene encoding chitin synthase 1 is essential for repairing damage to the intercellular septum incurred during the separation of daughter cells. The product of the chitin synthase 2 gene is specifically involved in the biosynthesis of the septum itself while that of the chitin synthase 3 gene produces most of the chitin in the bud scar and lateral cell wall. No single synthase appears to be essential for cell viability, but the loss of all three in mutants of *Saccharomyces cerevisiae* is lethal. Chitin synthase exists in multuple isozymic forms in *Candida albicans* and possibly in other pathogenic fungi.

The susceptibility of fungi to polyoxins and nikkomycins varies considerably and may be due to differences in the distribution and sensitivities of the chitin synthase isoforms to these antibiotics. Another factor which determines susceptibility to polyoxins is their transport into fungal cells by a permease that normally carries dipeptides. *Candida albicans* is intrinsically resistant to polyoxins because of the low activity of this permease. Despite these potential problems, it is hoped that the best of the currently available inhibitors of chitin synthase, nikkomycin Z, may eventually find a place in clinical medicine. The essential role of chitin in fungi has encouraged a search for other, more effective inhibitors of the chitin synthases, so far

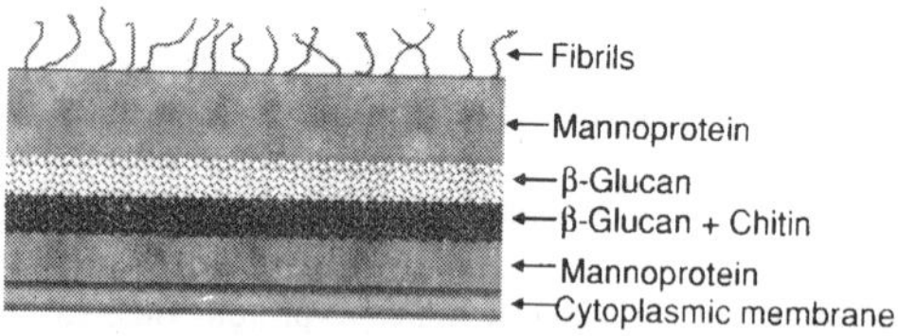

FIGURE 2.18 The general arrangement of layers in the fungal cell envelope. The components are not drawn to scale. It should be remembered that the precise structure of the fungal cell envelope is markedly species-specific.

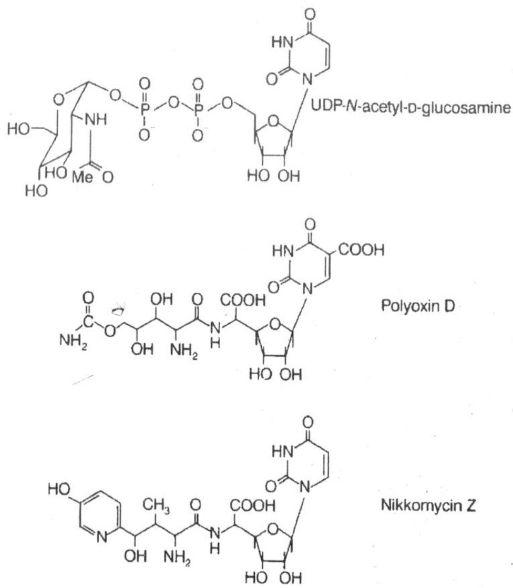

FIGURE 2.19 Antifungal agents that inhibit cell wall chitin synthesis, together with the substrate UDP-*N*-acetylglucosamine.

it must be said, with little success. Purification of the enzymes and the provision of adequate amounts for screening purposes are likely to be prerequisites for further progress.

2.6.2 Inhibitors of glucan biosynthesis

In glucan biosynthesis, the enzyme 1,3-β-glucan synthase catalyzes the transfer of glucose from UDP-glucose to the insoluble, growing glucan polymer. Unlike chitin synthase, glucan synthase has been purified to homogeneity. The enzyme consists of two subunits, one of which is an integral membrane protein, molecular mass 215 kDa, with multiple transmembrane helices. The other subunit is a much smaller protein (20 kDa) that interacts with GTP-binding proteins and is only loosely associated with the cell membrane. The function of the smaller subunit is apparently to activate the catalytic activity of the membrane-bound protein through interaction with the GTP-binding protein complexes. Two closely homologous forms of glucan synthase have been identified in *Saccharomyces cerevisiae*, designated as FKS1 and FKS2. FKS1 is dominant during vegetative growth whereas FKS2 has an essential role in sporulation. Genomic analysis of *Saccharomyces cerevisiae* predicts a third possible glucan synthase, FKS3, although it remains uncharacterized at present. Sequence-related glucan synthases have been found in other yeasts and in filamentous fungi. Five genes apparently encoding glucan synthases have been identified in the genome sequence of *Candida albicans*.

Echinocandin B (Figure 2.20) is a member of a large family of naturally occurring and semisynthetically modified lipopeptide antibiotics which have

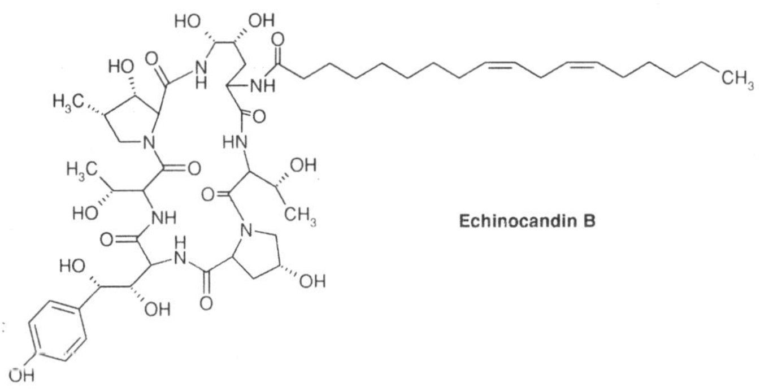

Echinocandin B

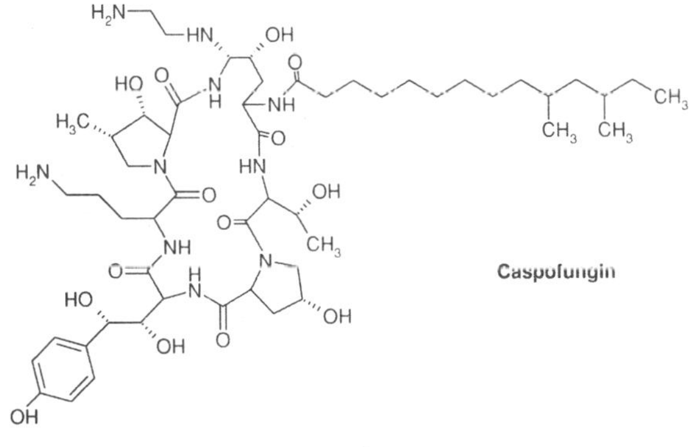

Caspofungin

FIGURE 2.20 Inhibitors of the biosynthesis of the glucan polymer in fungal cell walls.

potent activity *in vitro* against *Candida* spp. and against the filamentous *Aspergillus* spp. These compounds are powerful non-competitive inhibitors of 1,3-β-glucan synthase. This specificity may explain the lack of activity against fungi where glucan is not mainly 1,3-β-linked. In *Cryptococcus* spp., for example, a dangerous pathogen affecting the respiratory tract, the glucan is mostly 1,3-α-linked and the organisms are resistant to the cyclic lipopeptide antibiotics. Studies with an echinocandin-resistant mutant of *Saccharomyces cerevisiae* identified the membrane-bound component of 1,3-β-D-glucan synthase as the likely target of the drug. Unfortunately the details of the inhibitory mechanism, including the site of interaction between the antibiotic and the enzyme, are not known at present. The clinical usefulness of echinocandin B is limited by its propensity to cause lysis of red blood cells, which is thought to be due to its extended lipophilic side chain. Another member of the echinocandin group, Caspofungin (Figure 2.20), recently entered clinical practice for the treatment of *Aspergillus* infections unresponsive to other drugs and disseminated *Candida* infections. Caspofungin is likely to have the same mode of action as echinocandin B. Another semisynthetic cyclic lipopeptide, Micafungin, has recently enetered clinical practice in Japan.

2.6.3 Disruption of the function of mannoproteins

Pradimicin A (Figure 2.21) belongs to a unique group of antibiotics originally isolated from *Actinomadura hibisca* and is active against *Candida* spp., *Cryptococcus* spp. and *Aspergillus* spp. The antifungal action involves a change in the permeability of the cell membrane, which may result fom the ability of pradimicin to form an insoluble complex with mannan in the presence of calcium ions. Although this points to some form of interference with mannoprotein function, the biochemistry of the antifungal action of pradimicin requires further investigation. As yet, this drug has not been used to treat fungal infections in human patients although it has shown promise in the treatment of experimental infections.

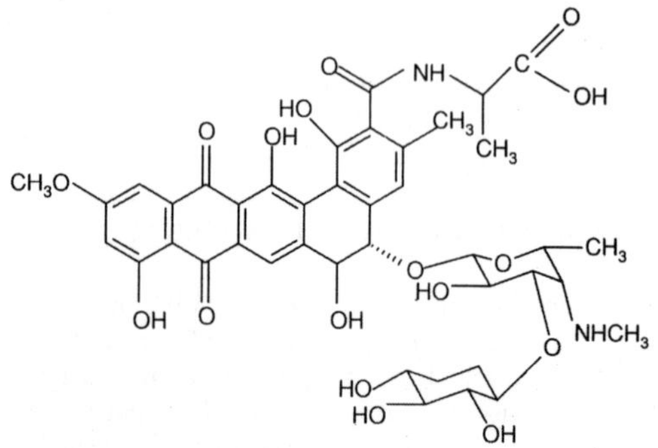

FIGURE 2.21 Pradimicin A, an experimental antibiotic active against several species of yeast pathogens. Its mode of action may depend upon interference with the function of cell wall mannoproteins.

Further reading

Allen, N. E. and Nicas, T. I. (2003). Mechanisms of action of oritavancin and related glycopeptide antibiotics. *FEMS Microbiol. Rev.* **26**, 511.

Brennan, P. J. (1995). The envelope of mycobacteria. *Annu. Rev. Biochem.* **64**, 29.

Denome, S.A. *et al.* (1999). *Escherichia coli* mutants lacking all possible combination of eight penicillin binding proteins: viability, characteristics and implications for peptidoglycan synthesis. *J. Bact.* **181**, 3981.

Doyle, R. J. and Marquis, R. E. (1994). Elastic, flexible peptidoglycan and bacterial cell wall properties. *Trends Microbiol.* **2**, 57.

Ghuysen, J.-M. *et al.* (1996). Penicillin and beyond: evolution, protein fold, multimodular polypeptides and multiprotein complexes. *Microbial Drug Resistance* **2**, 163.

Goffin, C. and Ghuysen, J.-M., (2002). Biochemistry and comparative genomics of SxxK superfamily acyltransferases. *Microbiol. Molec. Biol. Rev.* **66**, 702.

Green, D. W. (2002). The bacterial cell wall as a source of antibacterial targets. *Expert Opin. Ther. Targets* **6**, 1.

Liu, J. and Balasubramanian, M. K. (2001). 1,3β-Glucan synthase: a useful drug for antifungal drugs. *Curr. Drug Targets-Infect. Disord.* **1**, 159.

Ming, L.-J. and Epperson J. D. (2002). Metal binding and structure–activity relationship of the metallo-antibiotic peptide bacitracin. *J. Inorg. Biochem.* **91**, 46.

Odds F. C., Brown, A. J. P. and Gow, A. R. (2003). Antifungal agents: mechanisms of action. *Trends Microbiol.* **11**, 272.

Prescott, L. M., Harley, J. P. and Klein D. A. (1996). *Microbiology.* Wm. C. Brown, Dubuque, IA.

Schroeder E. K. *et al.* (2002). Drugs that inhibit mycolic acid biosynthesis in *Mycobacterium tuberculosis. Curr. Pharm. Biotechnol.* **3**, 197.

Skarzynski, T. *et al.* (1996). Structure of UDP-*N*-acetyl glucosamine enol pyruvyl transferase: an enzyme essential for the synthesis of bacterial peptidoglycan, complexed with substrate UDP-*N*-acetyl glucosamine and the drug fosfomycin. *Structure* **4**, 1465.

Van Hiejenoort, J. (2001). Formation of glycan chains in the synthesis of bacterial peptidoglycan. *Glycobiol.* **11**, 25R.

Chapter three

Antimicrobial agents and cell membranes

3.1 Microbe killers: antiseptics and disinfectants

The major interest throughout this book lies in the mechanism of action of drugs that can be used against microbial infections. For this purpose the compound must normally be absorbed and circulate in the blood. However, there is also a requirement in medicine, in industry and in the home for substances that kill bacteria and other micro-organisms on the surface of the body or in other places. Such products are known as disinfectants, sterilants, antiseptics or biocides, the choice of term depending on the circumstances in which they are used. 'Disinfectant' describes products intended for use in the presence of dirt and dense bacterial populations, for example, in cleaning animal quarters or drains. 'Biocide' is used more particularly for preservatives that prevent bacterial and fungal attack on wood, paper, textiles and other kinds of organic material and in pharmaceutical preparations. 'Antiseptic' is a term usually reserved for a substance that can be safely applied to the skin and mucosal surfaces to reduce the chances of infection by killing the surface bacteria. 'Sterilants' are substances used to sterilize an enclosed space; since penetration is paramount in this application, sterilants are usually gaseous.

The requirements for a compound having disinfectant or antiseptic action differ markedly from those needed in a systemic drug. Many compounds used successfully against microbial infections do not actually kill micro-organisms, but only prevent their multiplication, and most are inactive against nongrowing organisms. A cessation of microbial growth is often all that is needed in treating an infection, provided that the immune defences of the body can be mobilized to remove pathogens present in relatively small numbers. Furthermore, systemic antimicrobial agents often have a fairly limited spectrum of action. This is acceptable since the compound can be selected according to the nature of the infection that is being treated. Antiseptics and disinfectants, in contrast, are usually required to have a broad-spectrum killing effect. Antiseptics and preservatives used in ointments, creams, eyedrops and multidose injections must obviously be free from toxicity against the host tissues.

A distinction is often made between 'static' and 'cidal' compounds, but the division is by no means clear-cut. There is no certain way of determining whether a micro-organism is dead. The usual method of assessing the killing effect of an antiseptic is by measuring the 'viable count' of a previously treated bacterial or fungal suspension. The antiseptic is first inactivated and dilutions of the suspension are added to a rich medium. The organisms are deemed to be alive if they give rise to colonies. Many compounds are static at low concentrations and cidal at higher concentrations, and the effect may also depend on the conditions of culture. However, for antiseptics and disinfectants, a cidal effect is required under all normal conditions of application. Such compounds must be able to kill micro-organisms whether they are growing or resting; they must be able to deal with most of the common bacteria likely to be found in the environment and, ideally, fungi and viruses as well. Bacterial and fungal spores are usually much more difficult to kill.

Many of the older disinfectants are compounds of considerable chemical reactivity. Their antimicrobial action presumably depends on their ability to react chemically with various groups on or in the organism, thus killing them. Such compounds include hydrogen peroxide, the halogens and hypochlorites, the gaseous sterilants ethylene oxide, ozone, etc. Salts and other derivatives of the heavy metals, particularly of mercury, probably owe their antimicrobial effect to reaction with vital thiol groups. In disinfection, their high reactivity and toxicity limit their scope and they are not generally acceptable for the more delicate uses as antiseptics. For this purpose three main groups of compounds are widely used: alcohols, phenols and cationic antiseptics. The ether compound triclosan is also a popular and effective antiseptic. The main emphasis with these agents has been their efficacy against bacteria. Increasingly, however, there is concern that they should have useful activity against fungi and viruses. Although there are differences among the actions of the various types of antiseptics, they have several common features:

1. Antiseptics interact readily with bacteria, the amount adsorbed increasing with an increasing concentration in solution. The adsorption isotherm sometimes shows a point of inflection which corresponds to the minimum bactericidal concentration; higher concentrations lead to a much greater adsorption of the compound.
2. The extent to which micro-organisms are killed is governed by three principal factors: concentration of the antiseptic, cell density, and time of contact.The adsorption of a given amount of the compound per cell leads to the killing of a definite fraction of the microbial population in a chosen time interval.
3. The lowest concentration of the antiseptic that causes the death of micro-organisms also brings about leakage of cytoplasmic constituents of low molecular weight. The most immediately observed effect is a loss of potassium ions. Leakage of nucleotides is often detected by the appearance in the medium of material having an optical absorption maximum at 260 nm. Some loss of cytoplasmic solutes is not in itself lethal, and cells that have been rendered leaky by low concentrations of an antiseptic will often grow normally if they are immediately washed and placed in a nutrient medium. The increased permeability is a sign of changes in the membrane which may be initially reversible but become irreversible on prolonged treatment.
4. The necessary characteristic of antiseptics is their biocidal action, but there is often a low and rather narrow concentration range in which their effect is biostatic. At these low concentrations, certain biochemical functions associated with the microbial membrane may be inhibited but may not necessarily lead to cell death.
5. In the presence of higher concentrations of antiseptic and after prolonged treatment, the compound usually penetrates the cell and brings about extensive disruption of normal cellular functions, including inhibition of macromolecular biosynthesis and eventually precipitation of intracellular proteins and nucleic acids.

The primary effect of these antiseptics on the cytoplasmic membrane is thus established beyond doubt, but secondary actions on cytoplasmic processes are less well defined and may vary from one compound to another. Examples of evidence of action for particular compounds will be given as illustrations.

3.1.1 Phenols

Crude mixtures of cresols solubilized by soap or alkali and originally introduced as Lysol are still used as rough disinfectants. They need to be applied at high concentrations and are irritant and toxic, but they kill bacteria, fungi and some viruses. For more refined applications as antiseptics, chlorinated cresols or xylenols (Figure 3.1) are commonly used since they are less toxic than the simpler phenols. In general, the

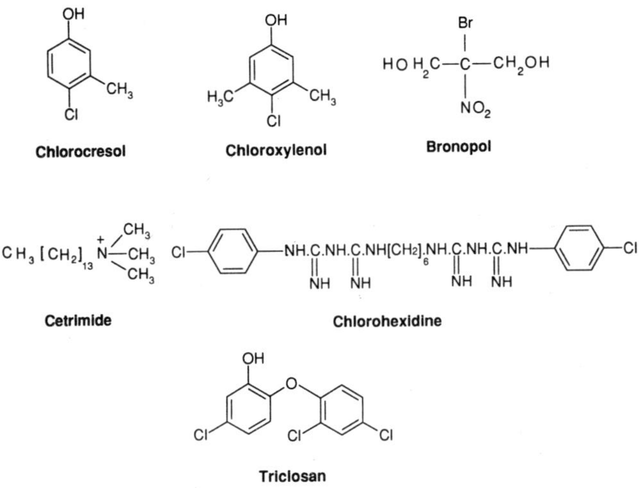

FIGURE 3.1 Some commonly used antiseptics. The formula for cetrimide shows the main components in the preparations normally sold. Homologues with other chain lengths, especially C_{16}, are also present.

primary action of the phenolic disinfectants and antiseptics is to cause the denaturation of microbial proteins, the first target being the proteins of the cell envelope, leading to lethal changes in membrane permeability.

3.1.2 Alcohols

Alcohols are widely used as inexpensive antiseptics, disinfectants and preservatives. Ethanol, for example, is a reasonably effective skin antiseptic as a 60–70% solution which kills both bacteria and viruses. Isopropanol (propan-2-ol) (at least 70%), which is slightly more effective as a bactericide than ethanol but is more toxic, can be used to sterilize instruments such as clinical thermometers. The more complex compound known as bronopol (2 bromo-2-nitropropane-1,3-diol, Figure 3.1) is an effective preservative for certain pharmaceutical products and toiletries, although there are concerns about the potential toxicity of some of the decomposition products of this compound, including formaldehyde and possibly nitrites, upon its exposure to light.

The antibacterial effects of the alcohols can be traced to a disruption of membrane function. The action of short-chain alcohols such as ethanol is probably dominated by the polar function of the hydroxyl group, which may form a hydrogen bond with the ester groups of membrane fatty acid residues. In contrast to ethanol, longer-chain alcohols gain access to the hydrophobic regions of membranes and this probably accounts for the increasing potency of antimicrobial action up to a maximum chain length of 10 carbon atoms. The interaction of alcohols with cell membranes produces a generalized increase in permeability which is lethal at higher concentrations. Bronopol may exert an additional effect by interacting with thiol groups in membrane proteins.

3.1.3 Cationic antiseptics

This classification covers a number of compounds differing considerably in chemical type. Their common features are the presence of strongly basic groups attached to a fairly massive lipophilic molecule. Although antiseptic action is found quite widely in compounds having these characteristics, the degree of activity is sharply dependent on structure within any particular group. For instance, in cetrimide, a quaternary alkylammonium compound (Figure 3.1), the length of the main alkyl chain is 14 carbon atoms and the activity of other compounds in the same series falls off markedly with longer or shorter chains. Cetrimide combines excellent detergent properties and minimal toxicity with a useful antiseptic action. However, it is not very potent against *Proteus* and *Pseudomonas* species and has little antiviral activity, except against viruses with a lipid envelope. Experiments with *Escherichia coli* labelled with ^{32}P show that with increasing concentrations of cetrimide the loss of cell viability closely parallels the degree of leakage of radioactivity from the bacteria. An effect on bacterial growth, however, is noticeable at concentrations that affect neither viability nor permeability. At bactericidal concentrations, the bacterial membrane is ruptured.

One of the best and most widely used of the cationic antiseptics is chlorhexidine (Figure 3.1). This compound has two strongly basic groups, both biguanides; it is often formulated as the digluconate, which has good solubility in water. Chlorhexidine is much less surface active than cetrimide and has little detergent action. However, it acts against a wide range of bacteria at concentrations between 10 and 50 $\mu g\ ml^{-1}$ and it also has useful activity against *Candida albicans*. Its toxicity is low and it has so little irritancy that it can be used on the most sensitive mucosal surfaces. For example, it is a useful aid to oral hygiene. Periodic rinsing of the mouth with chlorhexidine solution greatly reduces the population of *Streptococcus mutans* on the teeth. This minimizes the production of dental plaque and reduces periodontal infections that give rise to gingivitis. It also decreases the incidence of some types of caries. An important feature of this action is the strong binding of chlorhexidine to the tissues in the mouth, including the teeth, with subsequent slow release which maintains an antibacterial action over an extended period.

Chlorhexidine exerts effects on the cytoplasmic membrane that are characteristic of cationic antiseptics. At concentrations that just prevent the growth of *Streptococcus faecalis*, it inhibits the adenosine triphosphatase of the membrane. The effect can be shown in isolated membranes and on the solubilized enzyme derived from them. The proton motive force produced by the proton gradient across the cytoplasmic membrane, which drives ATP synthesis, is also dissipated. A similar concentration of chlorhexidine inhibits the net uptake of potassium ions by intact cells and suppresses macromolecular biosynthesis. The interaction between chlorhexidine and the cell membrane probably involves electrostatic binding between the cationic groups of the antiseptic and the anionic groups of phospholipids in the membrane. Hydrophobic interactions between the hexamethylene chains of chlorhexidine and the aliphatic chains of the phospholipids also contribute to the stability of the complex. When bacteria are treated with a range of concentrations of chlorhexidine and then examined for leakage of cytoplasmic solutes, the degree of leakage increases with concentration up to a maximum and then declines at higher concentrations. Low concentrations of chlorhexidine provoke the release of K^+ ions, nucleotides and sugars. Electron microscopy shows that the cells from higher levels of chlorhexidine treatment are grossly altered. The increased membrane permeability apparently allows the antiseptic to enter the cytoplasm and cause precipitation of the nucleic acids and proteins, resulting in the death of the cells. Under these circumstances, leakage is probably prevented by simple mechanical blockage. With Gram-negative bacteria, chlorhexidine damages the outer membrane as well as the cytoplasmic membrane. This can be seen as 'blistering' in electron micrographs (Figure 3.2). This phenomenon will be discussed further in connection with the action of polymyxin.

3.1.4 Triclosan: an antiseptic in a class of its own

Triclosan (2,4,4′-trichloro-2′-hydroxyphenyl ether, Figure 3.1) has been used for over 30 years as an antiseptic in medical handwashes and in many household products, including toothpaste, mouthwash, soaps, and deodorants, and in kitchen cutting boards. With broad-spectrum antimicrobial activity, although relatively poor against pseudomonads, and minimal toxicity (at least as a topical agent), triclosan has therefore enjoyed considerable success. For much of its history the antimicrobial activity of triclosan was thought to be caused by a direct membrane-damaging effect similar to that of the other common antiseptics. More recent research, however, reveals that the primary target of triclosan is lipid biosynthesis, the inhibition of which in turn leads to a loss of membrane integrity and func-

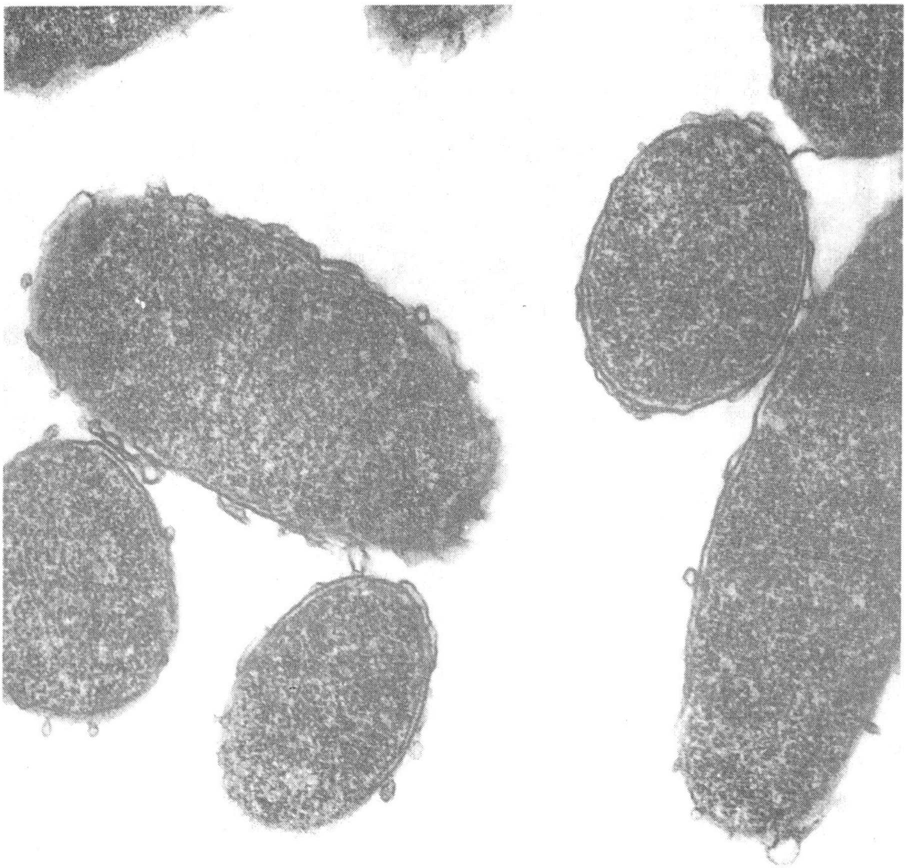

FIGURE 3.2 Electron micrograph of a cross-section of an *Escherichia coli* cell after treatment with chlorhexidine (30 μg ml^{-1}), showing 'blistering' of the cell wall.

tion. The first clue came with the isolation of a strain of *Escherichia coli* resistant to the antiseptic. Gene mapping located resistance in the *fabI* gene that encodes the enzyme enoyl-acyl carrier protein reductase, or FabI. This enzyme plays a key role in the biosynthesis of short-chain fatty acids and is a homologue of the InhA enzyme involved in mycolic acid biosynthesis in mycobacteria described in Chapter 2. Purified FabI from triclosan-resistant *Escherichia coli* was unaffected by the antiseptic, whereas the enzyme from the triclosan-sensitive bacterium was strongly inhibited. The essential cofactor for FabI is NAD^+ and, remarkably, triclosan increases the affinity of the enzyme for NAD^+ by some four orders of magnitude. A tight ternary complex is formed that involves FabI, NAD^+ and triclosan with a dissociation constant for triclosan of 20–40 pM. X-Ray analysis of the complex reveals that the 2-hydroxy-3-chlorophenyl ring of triclosan interacts with enyzme-bound NAD^+ and with the enzyme itself via hydrogen bonding to the phenolic hydroxyl group of tyrosine-156. The 2,4-dichlorophenyl ring of the antiseptic, which is rotated 90° out of the plane of the hydroxychlorophenyl ring, forms a hydrogen bond between the 4-Cl atom and alanine-95. There are also hydrophobic contacts between this ring and the side chain of methionine-159.

It is intriguing that the InhA enzyme of *Mycobacterium tuberculosis,* which is the molecular target of the active metabolite of the antitubercular drug isoniazid (Chapter 2), is also inhibited by triclosan. However, *Mycobacterium tuberculosis* is not susceptible to triclosan and there is evidence that the bacterium harbors another form of enoyl-aCP reductase which is not inhibited by triclosan and therefore assumes the function of InhA when that enzyme is inhibited. These observations, along with the recent finding that systemic doses of triclosan are effective against experimental infections in mice, suggest the possibility that the structure of triclosan could conceivably provide a starting point for the design of novel antitubercular drugs.

3.2 Cationic peptide antibiotics

Several classes of polypeptide antibiotics are known, all of which have their origins in natural sources. Compounds with cyclized peptide regions include the tyrocidins, gramicidins and polymyxins (Figure 3.3). The tyrocidins and gramicidin S are cyclic decapeptides. These contain one or sometimes two free amino groups. They are more active against Gram-positive than against Gram-negative bacteria. The polymyxins have a smaller polypeptide ring attached to a polypeptide chain terminating with a branched 8- or 9-carbon fatty acid residue. They have five free amino groups associated with the diaminobutyric acid units. All the cyclized peptide antibiotics carry a positive charge. The antibacterial action of polymyxins is directed particularly against Gram-negative organisms although the selectivity can be altered dramatically by chemical modification. For example, the penta-*N*-benzyl derivative of polymyxin is highly active against Gram-positive bacteria.

In addition to the cyclized peptide antibiotics, there are several hundred linear peptides with antimicrobial activity which have been isolated from many different species of prokaryotes and eukaryotes. These linear peptides, e.g. magainin II (Figure 3.3), have an enormous variety of sequences and structures but retain certain features in common. They are generally 12–50 amino acids in length and, like the cyclized peptides, carry an overall net positive charge owing to an excess of basic lysine and arginine residues over acidic amino acids.

Polypeptide antibiotics have only a minor place in medicine because they also damage mammalian cell membranes. The polymyxins may be used systemically in severe *Pseudomonas* infections, although there is considerable risk of kidney damage. Polymyxin is bactericidal and acts on nongrowing as well as growing cells. At low concentrations its bactericidal action parallels the degree of release of cytoplasmic solutes. It is strongly and rapidly bound to bacteria.

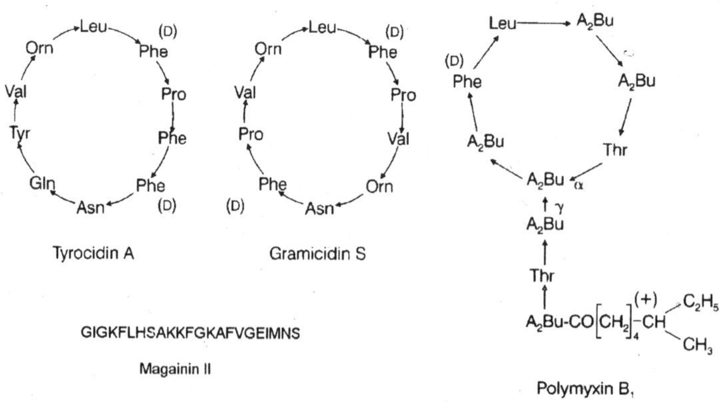

FIGURE 3.3 Peptide antibiotics that damage bacterial cell membranes. A_2Bu: 2,4-diaminobutyric acid. The arrows show the direction of the peptide bonds. Except where shown, all peptide linkages involve α-amino and α-carboxyl groups. Configurations are L unless otherwise indicated. The formula for magainin II uses the conventional abbreviations for amino acids.

With *Salmonella typhimurium*, the binding of 2×10^5 molecules of polymyxin per cell is known to be bactericidal. In Gram-negative bacteria, antibiotics of the polymyxin group apparently bind first to the outer membrane, affecting mainly the lipopolysaccharide. The gross effects of polymyxin on the outer membrane are sometimes revealed in electron micrographs as blisters, similar to those caused by chlorhexidine (Figure 3.2). The swellings may be due to an increase in the surface area of the outer leaf of the outer membrane. The parallels between the action of polymyxin and chlorhexidine are quite striking. In both, the binding and antibacterial effects are antagonized by excess of calcium or magnesium ions, indicating that the displacement of divalent ions is an important feature of their action. The disorganization of the outer membrane by polymyxin enables the antibiotic to gain access to the cytoplasmic membrane, which in turn is damaged by the antibiotic. The disruption of normal membrane function brought about by polymyxin results in a generalized increase in membrane permeability and the loss of essential cellular nutrients and ions, such as K^+.

Physical measurements of various kinds all support the conclusion that the antibacterial action of polymyxin is caused primarily by its binding to membranes. The positively charged peptide ring is thought to bind electrostatically to the anionic phosphate head groups of the membrane phospholipid, displacing magnesium ions which normally contribute to membrane stability. At the same time, the fatty acid side chain is inserted into the hydrophobic inner region of the membrane. The effect is to disturb the normal organization of the membrane and to alter its permeability characteristics.

The tyrocidins (Figure 3.3) are also bactericidal and promote leakage of cytoplasmic solutes. Their action on the bacterial membrane permits passage into the cell of ions that are normally excluded, and under some conditions this causes uncoupling of oxidative phosphorylation as a secondary effect. Gramicidin S, a closely related compound (Figure 3.3), acts similarly. It lyses protoplasts from *Micrococcus lysodeikticus* but not those from *Bacillus brevis*. Since it is bactericidal towards the former organism but not the latter, it is reasonable to suppose that both its action and specificity depend upon its effect on the cytoplasmic membrane, but a detailed explanation is lacking. The tyrocidins act not only on bacteria but also on the fungus *Neurospora crassa*. In this organism, concentrations of the antibiotic that stop growth and cause leakage of cell contents also cause an immediate fall in membrane potential, a consequence of the destruction of the permeability barrier.

In both the tyrocidin group and in the polymyxins, the cyclic molecular structure is important for antibacterial activity. The presence of basic groups is also essential, but in other respects the molecules can be varied considerably without losing activity. The simple symmetrical structure of gramicidin S has been subjected to many modifications. Activity is preserved when the ornithine units are replaced by arginine or lysine groups but is lost by modifications that destroy the basic character of the terminal groups. The compound in which glycine replaces L-proline is fully active. Moreover one L-proline residue together with the adjacent D-phenylalanine can be replaced by a δ-aminopentanoic acid group without losing antibacterial activity. The resulting compound has only nine peptide groups, but retains the same ring size. Acyclic compounds having the same sequence of amino acids as gramicidin S show only slight antibacterial action.

The importance of the cyclic structure lies in the maintenance of a well-defined, compact conformation in solution. This has been shown by nuclear magnetic resonance, optical rotatory dispersion and other physical measurements. In tyrocidin A and gramicidin S the conformation is determined by lipophilic association between the nonpolar side chains of the amino acids, particularly leucine, valine, proline and phenylalanine, and by hydrogen bonding between the peptide groups. Three regions have been defined in the molecular topography of tyrocidin A: a hydrophobic surface; a flat hydrophilic opposite surface consisting of the peptide groups of most of the amino acids in equatorial positions; and a helical hydrophilic region, accommodating the amide groups of asparagine and glutamine and the tyrosine hydroxyl group. Gramicidin S shows a similar arrangement, based on a pleated-sheet structure. In both antibiotics the ornithine amino groups, which are essential for antibacterial activity, stand out from the hydrophilic surface.

Out of the enormous variety of linear cationic peptides with antibiotic activity, only a handful of compounds have reached the stage of clinical evaluation. The most interesting to date appear to be the magainins (Figure 3.3), which were originally isolated from frog skin. The action of the linear cationic peptides on micro-organisms is broadly similar to that of the cyclized compounds. That is, in the case of Gram-negative bacteria, there is an initial interaction with the negatively charged lipopolysaccharides on the the surface of the outer membrane. This enables the peptides to insert into and cross the outer membrane. In the case of Gram-positive bacteria and fungi, the thick cell wall provides only a partial barrier to the cationic peptides. The cytoplasmic membrane is then open to attack. The linear peptides again interact electrostatically with the anionic outer surface of the membrane, allowing the peptides to insert into the lipid bilayer. Precisely what happens to the membrane at this point is unclear; one suggestion is that the peptides may orient themselves to form channels across the membrane in the manner of the staves of a barrel, the so-called barrel-stave model. Alternatively, the peptides may bind so effectively and in such quantity to the outer surface of the cytoplasmic membrane that they wreck the intergrity of the membrane entirely. This is known as the carpet model. Whatever the exact details, the fundamental permeability characteristics of the cytoplasmic membrane are compromised, allowing the loss of essential solutes from the cytoplasm and the possible ingress of peptide antibiotics into the cytoplasm, with subsequent disruptive interactions with vital macromolecular entities such as enzymes and other proteins.

3.3 Ionophoric antibiotics

Several classes of antibiotics may be grouped together because of their common property of facilitating the passage of inorganic cations across membranes by the formation of hydrophobic complexes with the ions or by forming ion-permeable pores across the membranes. Although these compounds were discovered through their antibacterial activity, they are not used in human bacterial infections because of their lack of specificity. They act equally effectively on the membranes of animal cells and may therefore be toxic. However, some ionophores have applications in veterinary medicine and animal husbandry which are discussed later. Ionophoric compounds are also of considerable biochemical interest and are widely used as experimental tools. As antimicrobial agents, they are active against Gram-positive bacteria whereas Gram-negative bacteria are relatively insensitive because their outer membranes are impermeable to hydrophobic compounds of the molecular size of the ionophores. Monensin and lasalocid have useful activity against the protozoal parasite *Eimeria tenella,* the organism that causes coccidiosis in poultry and against *Clostridium perfringens,* another pathogen of economic importance in the broiler chicken industry. There is also some evidence for the selective toxicity of certain ionophores against the malarial parasite (see below).

3.3.1 Valinomycin

This was the first member of a group of related compounds to be discovered and is among the most widely studied. It is a cyclic depsipeptide in which amino acids alternate with hydroxy acids in a ring that contains both peptide and ester groups (Figure 3.4). An important feature, which is common to all the cyclic ionophores, is the alternation of D- and L-configurations in pairs around the 12 components of the ring structure.

Valinomycin forms a well-defined complex with potassium ions. X-Ray analysis of this complex reveals a highly ordered structure (Figure 3.5) in which the potassium atom is surrounded by six oxygen atoms. The ring structure is puckered and held in a cylindrical or bracelet-like form by hydrogen bonds roughly parallel to its axis. The ability to achieve such a conformation depends entirely on the alternation of D- and L-centres. The dimensions are such that the potassium atom is exactly accommodated. The ion entering the complex must shed its normal hydration shell; the complex retains the positive charge carried by the ion. The structure observed in the crystal is substantially maintained in solution. Although valinomycin will also form a complex with sodium, the smaller sodium atom fits much less exactly into the structure, and this complex has a stability constant 1000 times smaller than that of the potassium complex.

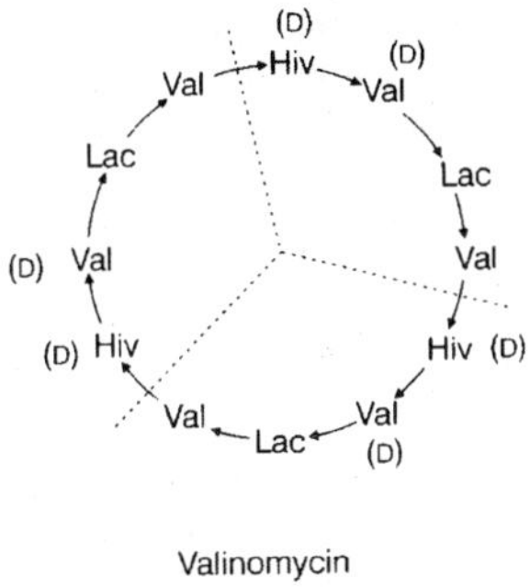

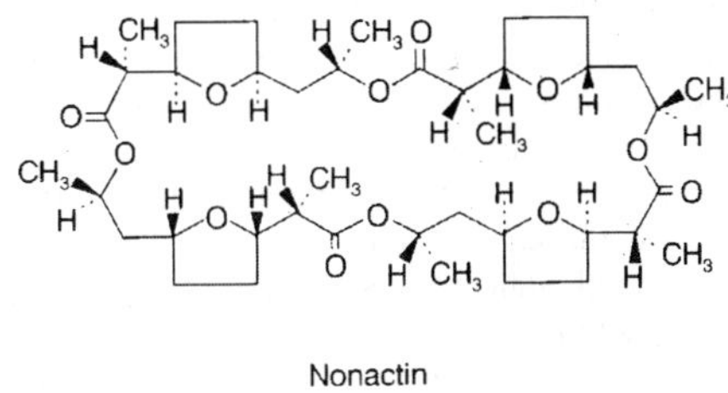

FIGURE 3.4 Antibiotics enhancing the permeability of membranes to potassium ions. In the valinomycin structure; Val represents valine; Lac, lactic acid; and Hiv, 2-hydroxyisovaleric acid. The arrows indicate the direction of peptide or ester bonds. Configurations are L unless otherwise indicated. Dotted lines separate the repeating units.

The high specificity of valinomycin towards the potassium ion and the physical properties of the complex are consistent with its postulated action on biological membranes. The binding of the potassium ion in the structure of valinomycin increases the lipophilicity of the antibiotic and thereby promotes its diffusion into the hydrophobic regions of the membrane. The lipophilic molecule moves physically through the membrane lipids, carrying potassium, and returns in the protonated form. In a passive membrane the flow is determined solely by the concentration of potassium ions on each side of the membrane, but in mitochon-

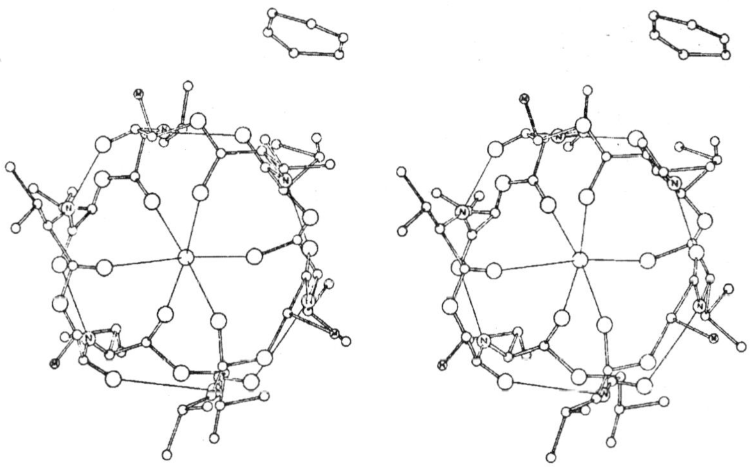

FIGURE 3.5 Stereophotographs of a model of the potassium complex of valinomycin. To obtain a three-dimensional effect, the diagram should be held about 50 cm from the eyes and attention concentrated on the space between the two pictures. With practice, three pictures can be seen, the middle one showing a full stereoscopic effect. The central metal ion is seen coordinated to six oxygen atoms. Nitrogen atoms are labelled N and the methyl groups of the lactyl residues are labeled M. Hydrogen bonds are shown by thin lines. The solitary hexagonal ring is a hexane of crystallization. (We are grateful to Mary Pinkerton and L. K. Steinrauf for allowing us to reproduce this picture.)

dria supplied with an energy source, potassium is taken in by an energy-coupled process against the concentration gradient. The process is highly effective, one valinomycin molecule being able to transport 10^4 ions per second, a turnover rate higher than that of many enzymes. The transport of potassium by valinomycin and similar ionophores shows saturation kinetics with respect to the cation; sodium ions inhibit potassium transport although they undergo little transport themselves. The kinetic results are well explained by a model in which the ionophore at the membrane surface first forms a hydrophilic cation complex. This is transformed into a hydrophobic complex which can then cross the membrane. The rate of the transformation from one type of complex to the other determines the turnover number.

Valinomycin specifically drains Gram-positive bacteria of potassium and growth ceases because of the requirement for potassium in cellular metabolism. If the potassium content of the medium is raised to that normally present in the cytoplasm, the inhibitory action of valinomycin is prevented. A characteristic secondary effect of valinomycin in growing aerobic bacteria is to disturb oxidative phosphorylation. In analogous fashion, valinomycin disrupts oxidative phosphorylation in the mitochondria of eukaryotic cells.

3.3.2 Nonactin

Another series of antibiotics known as the macrotetrolides, exemplified by nonactin (Figure 3.4), have a cyclic structure which also permits the enclosure of a potassium ion in a cage of eight oxygen atoms (the carbonyl and tetrahydrofuran oxygens), with the rest of the molecule forming an outer lipophilic shell. To produce this structure, the ligand is folded in a form resembling the seam of a tennis ball and is held in shape by hydrogen bonding. The action of the macrotetrolides closely resembles that of valinomycin.

3.3.3 Monensin

Ionophoric antibiotics of another broad group, typified by monensin (Figure 3.6), carry a carboxyl group. In these compounds the molecule itself is not cyclic, but as with valinomycin, a metal complex is formed in which the ion is surrounded by ether oxygen atoms and the outer surface is lipophilic. At physiological pH the carboxylate group of monensin is likely to be ionized and located in the external environment of the cell. The lipophilic character of the rest of the molecule enables it to insert into the lipid bilayer of the cell membrane. Positively charged, monovalent metal ions in the aqueous medium associate initially with the negatively charged carboxyl ions of monensin molecules, converting them into neutral species. Water molecules of solvation bound to free metal ions are lost during the interaction with the antibiotic. NMR spectroscopy shows that following the initial interaction of a metal ion with the carboxyl group, the monensin molecule wraps itself around the metal, with five oxygen atoms binding strongly and a sixth oxygen interacting more weakly with the enclosed metal atom. This folding of the molecule brings the carboxyl group at one end into a position where it can form strong hydrogen bonds with the alcohol groups at the other end; the structure is thus stabilized into an effectively cyclic form. Sodium ions are bound by monensin in preference to potassium ions. The presence of the carboxyl group promotes an electrically neutral cation–proton exchange across the membrane by moving as an undissociated acid in one direction and as a cation–anion complex with no net charge in the other direction. This distinguishes monensin and stucturally similar ionophores from valinomycin and nonactin, in which the metal complex carries a positive charge. The mechanism by which the sodium ion is disgorged from the monensin complex at the opposite face of the membrane seems likely to involve a sequential reversal of the association process as the complex encounters the internal aqueous environment.

Monensin is a compound of considerable commercial importance. It was first introduced as a treatment for the protozoal infection cocciodiosis in chickens, and proved to be of exceptional utility. It has shown few signs of the development of resistance which usually limits the effective life of drugs sold for treating coccidiosis. Monensin also improves the utilization of feedstuffs in ruminants. Its action depends on altering the balance of free fatty acid production by rumen bacteria in favour of propionate at the expense of acetate. Propionate is energetically more useful to

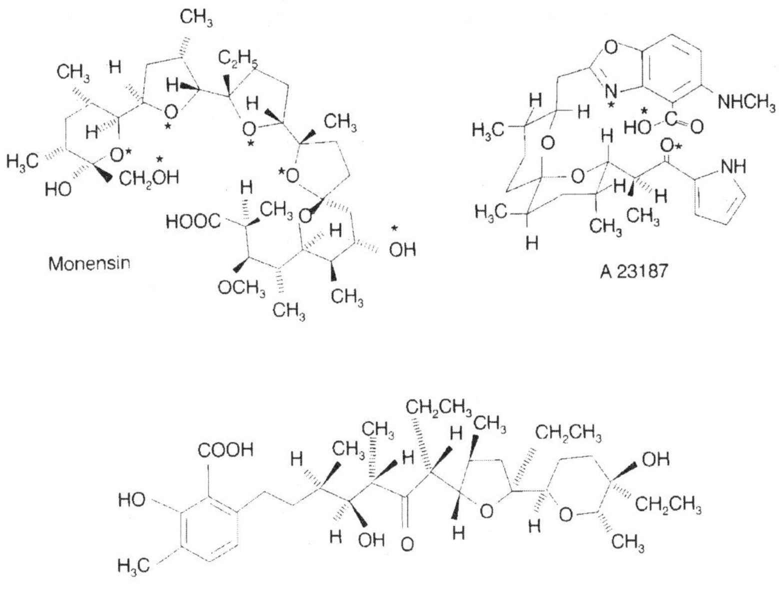

FIGURE 3.6 Three more ionophoric antibiotics. Monensin preferentially complexes sodium ions by coordination with the oxygen atoms marked by asterisks. The other two compounds complex with calcium ions. All three compounds have coccidiostatic activity.

the animal than acetate. There is also a lessening of the metabolically wasteful production of methane. The molecular basis of these effects is not certain, but the shifts in rumen metabolism can probably be attributed to differential antimicrobial actions on the complex population of micro-organisms in the rumen. The action of monensin on cell membranes is not species-specific. Its lack of toxicity when given orally to farm animals probably depends upon its limited absorption from the gastrointestinal tract.

3.3.4 Ionophoric antibiotics specific for divalent cations

The ionophores considered so far form complexes only with monovalent metal ions. A few ionophores are known that form complexes with divalent ions. One of the best known of these is the antibiotic calcimycin, otherwise known as A23187 (Figure 3.6). This forms a 2:1 complex with calcium or magnesium ions, the calcium complex having the higher stability; it binds monovalent ions only weakly. As with monensin, it is not a cyclic molecule but is able to fold into an effectively cyclic conformation by the formation of a hydrogen bond between a carboxyl oxygen and the NH group of the pyrrole ring. The divalent metal ion is held in octahedral co-ordination between the polar faces of two ligand molecules. This gives an electrically neutral complex with a hydrophobic outer surface. It acts as a freely mobile carrier of these ions and causes progressive release of magnesium, uncoupling of oxidative phosphorylation and inhibition of adenosine triphosphatase in mitochondria suspended in a magnesium-free medium. Like monensin, divalent cationophoric antibiotics have coccidiostatic activity, for example, the antibiotic lasalocid, or X-537A (Figure 3.6).

Gramicidin A

Gramicidin A (Figure 3.7) (which is quite unrelated to gramicidin S) has many biochemical properties resembling those of valinomycin. It shows a specificity towards potassium ions and promotes their passage across lipid membranes. However, studies have shown that its mechanism of action is different. The most significant demonstration of this distinction depends upon measurements of the electrical conductivity of artificial membranes separating aqueous layers containing potassium ions. Conditions can be chosen where addition of valinomycin, nonactin or gramicidin A at 0.1 μM concentration lowers the resistance of the

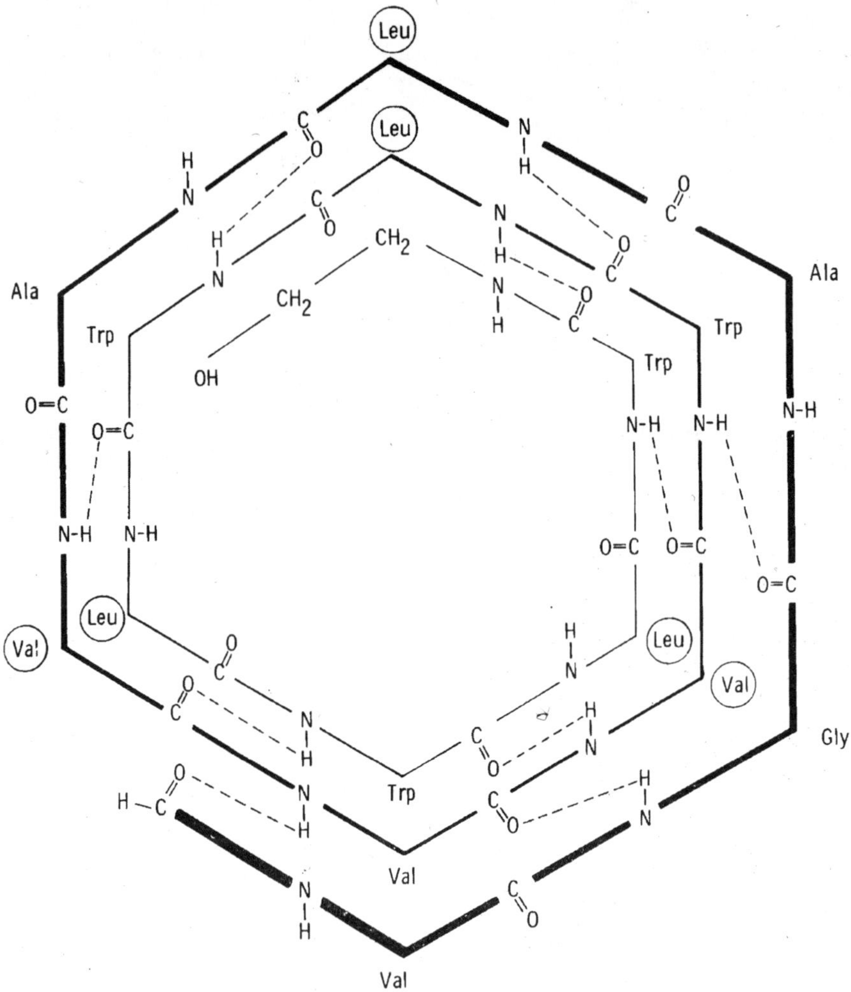

FIGURE 3.7 Gramicidin A. One possible helical structure having 6.3 residues per turn. Bonds drawn inwards are directed down the helix; those drawn outwards are directed up. D-Amino acid residues are circled.

membrane at least 1000-fold. If the temperature is then lowered gradually, the membrane reaches a transition point at which its lipid layer effectively changes phase from liquid to solid. In the presence of valinomycin or nonactin, a 2° C fall in temperature at the transition point causes a dramatic rise in membrane resistance, but in a similar experiment with gramicidin A, resistance rises only slowly as the temperature falls. The effect with compounds of the valinomycin type is understandable since they require a liquid membrane for mobility and movement. Gramicidin A acts by the formation of a pore that permits the flow of ions through a rigid membrane. Gramicidin A is a linear polypeptide in which alternating amino acid residues have the L-configuration. The remaining residues are either D-amino acids or glycine. The C-terminal is amidated with ethanolamine and the N-terminal carries a formyl group. The configuration allows the molecule to form an open helical structure held together by hydrogen bonds lying almost parallel to the axis of the cylinder. One possible helical form is shown in Figure 3.7. The inside of the helix is lined with polar groups and there is a central hole about 0.4 nm in diameter. The fatty side chains of the amino acids form a lipophilic shell on the outside. One such molecule is not long enough to form a pore across a membrane, but head-to-head dimerization is believed to occur by bonds between the formyl groups. The existence of dimerization is supported by measurements in artificial membranes which show that conductance is proportional to the square of the concentration of gramicidin A. The length of the dimer is calculated to be 2.5–3.0 nm, which is somewhat less than the thickness of the fatty acid layer in many membranes, so some distortion probably occurs during pore formation.

Conductivity measurements suggest that these pores have a transient existence, a small fraction of the antibiotic being in the form of pores at any given time. The life of a channel measured in a phosphatidylethanolamine artificial membrane was only 0.35 s. However, while a pore is in existence, its transporting capacity is high. One channel is estimated to convey 3×10^7 K^+ ions s^{-1} under a potential gradient of 100 mV. Thus a low concentration of gramicidin A is a very effective carrier of potassium ions. Divalent cations are too large to traverse the gramicidin pores but block the free passage of monovalent ions.

3.3.5 Antiprotozoal activity of ionophores

With the exception of monensin and lasalocid, the practical applications of ionophores are very limited. However, the pore-forming ionophores gramicidin D, a linear peptide structurally related to gramicidin A, and lasalocid show a surprising degree of selective toxicity for the malarial protozoal parasite *Plasmodium falciparum*. The sensitivity of the organism was studied during its intraerythrocytic stage. The normal permeability characteristics of the red cell membrane are drastically altered by the intracellular presence of the parasite; permeability to Na^+ and Ca^{2+} ions is increased whereas permeability to K^+ is decreased. The selective toxicity of the ionophores may be due to preferential partitioning into the cell membranes of infected red cells, leading to abnormal ion fluxes into the intraerythrocytic environment of the malarial protozoan. It remains to be seen whether the interesting *in vitro* activities of the ionophores can be translated into safe and effective treatments for malarial infection.

3.4 Antifungal agents that interfere with the function and biosynthesis of membrane sterols

3.4.1 Polyene antibiotics

The polyenes constitute a large group with varied molecular structures which interact with membranes in an especially interesting way. There are about 200 polyenes, all produced by *Streptomyces* spp. Of these, only a few are sufficiently nontoxic for clinical use and only one is used to treat systemic infections in man—amphotericin B (Figure 3.8). Polyenes are active against fungi, including yeasts, but not against bacteria. Systemic fungal infections are potentially dangerous, and intravenous amphotericin B, which is usually specially formulated with lipids, can halt infections which might otherwise be fatal. Nystatin (Figure 3.8) is useful as a topical agent to treat localized candidal infections on mucosal surfaces. The polyenes are not absorbed from the gastrointestinal tract but can be given by mouth to combat fungal growth in the oral cavity and gut. The relatively low toxicity of pimaricin (Figure 3.8)

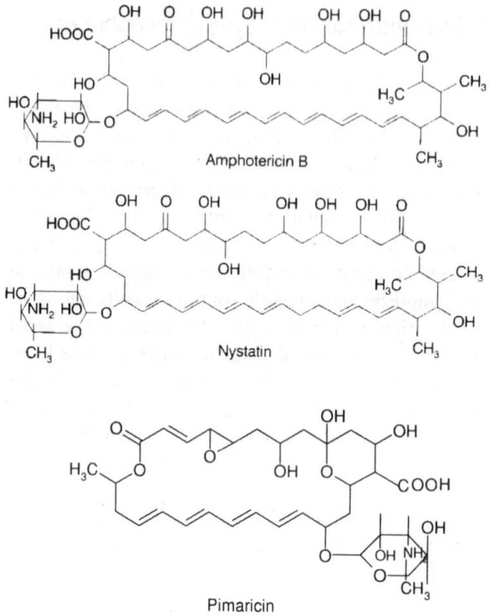

FIGURE 3.8 Polyene antifungal agents.

permits its use in food as a preservative against fungal contamination.

The primary site of interaction of polyenes is the fungal sterol, ergosterol. The sterol composition of the fungal membrane is important in determining the sensitivity to the polyene antibiotics, and the difference in their relative affinities for ergosterol (the major sterol of fungal membranes) and the predominant cholesterol of mammalian membranes allows limited clinical applications of the polyenes. The ergosterol molecule adopts a cylindrical three-dimensional structure which may favour its interaction with polyenes in preference to the sigmoidal structure of cholesterol. Although the molecular details of the interactions between polyenes and ergosterol and cholesterol remain uncertain, the similarity between the sterols means that the therapeutic safety margins of polyenes are low. The compounds are therefore inherently toxic to mammals, and the side effects, such as kidney and nerve damage, of even the best compound, amphotericin B, restrict their clinical use. This toxicity can be reduced to some extent by administering the drug as a complex with cholesterol and phosphatidylglycerol.

Bacteria are not affected by amphotericin B, or other polyenes since their membranes do not contain sterols. The mycoplasma *Acholeplasma laidlawii* does not need sterols in its membrane, but can incorporate either cholesterol or ergosterol when they are added to the growth medium. Ergosterol-containing organisms are sensitive to amphotericin B, whereas cells grown without sterols are not affected. The addition of ergosterol or digitonin to yeast cell cultures prevents amphotericin from being toxic; this results from the complexing of amphotericin by these agents. Mutant yeasts which have a block at some stage of ergosterol synthesis are resistant to amphotericin because there is no longer a target for the polyene in the cell. Fortunately this type of resistance is not clinically important.

The general action of polyenes is to increase the permeability of fungal membranes, but the specific effects of individual polyenes show considerable differences. Filipin, for example, causes gross disruption of membranes, with release of K^+ ions, solutes of low molecular weight and small proteins, whereas *N*-succinylperimycin more specifically induces the release of intracellular K^+ ions. Living cells cannot survive a catastrophic loss of intracellular potassium, so the interaction of polyenes with cell membranes soon results in cell death. The ion permeability-enhancing effects of the polyenes are probably caused primarily by the drug molecules creating pores in the membranes. Molecular models of amphotericin B and nystatin show a rod-like structure held rigid by the all-*trans* extended conjugated system, which is equal in length to an ergosterol molecule. The cross-section of the polyene structure is roughly rectangular. One surface of the rod is hydrophobic and the opposite surface, studded with axial hydroxyl groups, is hydrophilic. At one end of the rod the mycosamine sugar group and the carboxyl group form a zwitterionic assembly with strongly polar properties.

A computer-based, 'virtual' model of a possible pore or channel involves eight amphotericin B and eight cholesterol molecules (the latter molecule was chosen because of the investigators' interest in polyene toxicity). The computer also simulated the environment of the membrane with layers of phospholipid surrounding the channel. The length of channel was less than the thickness of typical cytoplasmic membranes and it is possible that in reality two such channels end-to-end may actually bridge the membrane. It is also thought possible that the lipids around the channel may accommodate themselves so that the bilayer is 'pinched in' somewhat, with the bilayer now approximating the channel length. The stability of the computer model of the channel is largely dependent upon hydrogen bonding between the hydroxyl groups of neighbouring amphotericin B molecules and the amino and carboxyl groups of adjacent drug molecules. The hydroxyl groups line the internal surface of the channel and provide the necessary hydrophilic environment for the passage of K^+ and other water-soluble ionized species. It is surprising that the computer simulation did not reveal any structurally specific interactions between the sterol and amphotericin B, although the cholesterol molecules are essential to the formation of the channel. Further details of this fascinating, though speculative, approach to the puzzle of the amphotericin B membrane channels are available in a research paper included under 'Further reading'. Considerably more research will be needed before the precise details of these channels are revealed. Furthermore, the formation of pores does not satisfactorily explain the gross changes in permeability brought about by polyenes such as filipin. In this case it seems likely that the insertion of the antibiotic into membranes causes a more general disruption of their organization and function.

3.4.2 Inhibition of ergosterol biosynthesis

Whereas the polyenes disrupt membrane function through direct interaction with membrane ergosterol, there are several groups of antifungal compounds that act by inhibiting the biosynthesis of this sterol. An outline of the biosynthetic pathway from squalene to ergosterol is shown in Figure 3.9 and indicates the points of inhibition of several types of antifungal agents that are useful in either medicine or agriculture.

Azoles

The azoles are among the most important compounds currently in use against fungal infections. They are subdivided into imidazoles or triazoles, according to whether they have two or three nitrogen atoms in their five-membered azole ring (Figure 3.10). Unlike the polyenes, most azoles do not kill fungi but rather act as fungistatic agents. In their favour is that they are relatively nontoxic, with the exception of some of the older compounds, such as ketoconazole, which has caused fatal liver damage on rare occasions.

The antifungal action of the azoles depends on inhibition of the C-14 demethylation reaction in the biosynthesis of ergosterol. The enzyme involved, 14α-sterol demethylase, is a P_{450} cytochrome protein, otherwise known as P450-Erg11P or Cyp51p, according to different gene-based nomenclatures. The azoles inhibit the enzyme by forming a stoichiometric complex with the iron of the heme component of the enzyme. The formation of the complex is detected by the red shift of the Soret band of heme from 417 to 447 nm. The heme iron interacts with the lone pair of electrons on one of the ring nitrogens, and the complex is further stabilized by interactions between hydrophobic moieties in the azole ligand and the enzyme. The conformation of the active center of 14α-sterol demethylase varies among fungal species and the many mammalian P_{450} mono-oxygenases. The precise details of the interaction between each of the azole drugs and the various P_{450} enzymes are therefore likely to influence the inhibitory potency across different fungal species as well as the potential toxicity in patients (see later discussion). The azole-mediated inhibition of the demethylase is noncompetitive for the sterol substrate and leads to a greater net reduction in flow through the metabolic pathway than competitive inhibition. The result is an accumulation of methylated sterols in the cell and a reduction in the ergosterol content. Methylated sterols are more bulky than ergosterol and do not fit easily into a normal membrane structure. This interference in the membrane structure is thought to have adverse effects on membrane-bound enzymes, such as those concerned with chitin synthesis and nutrient uptake, either directly on their activity or on their control. The depletion of ergosterol may also result in interference with its hormone-like actions on cell growth.

Typical examples of azole antifungals include the topically active agent miconazole (Figure 3.10), which is effective against thrush and dermatophyte infec-

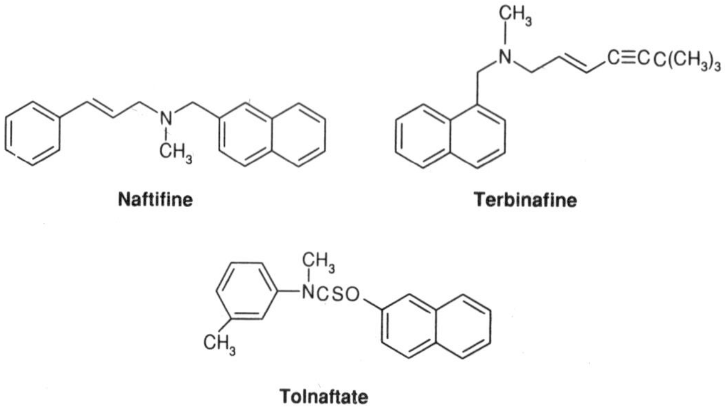

FIGURE 3.11 Allylamine antifungal agents used to treat dermatophytic infections.

of the ergosterol biosynthetic pathway. The first target is the enzyme that catalyzes the reduction of the double bond at the 14–15 position formed after the removal of the C-14 methyl group. The second target is the isomerization of the double bond between C-8 and C-9 of fecosterol to a position between C-8 and C-7 (Δ^7–Δ^8 isomerase). The balance between these two inhibitory activities varies from fungus to fungus and probably reflects subtle differences in the enzymes involved. Tridemorph (Figure 3.12) inhibits *Ustilago maydis* mainly at the C-14 reduction step, whereas *Botrytis cinerea* is inhibited mainly at the C-8 to C-7 isomerization. However, in general the more important target is likely to be the Δ^{14} reduction since this enzyme is essential for fungal viability whereas the Δ^7–Δ^8 isomerase is not. There are serious concerns over the safety of tridemorph, specifically in relation to an apparent link with birth defects. As a result, the use of the compound in agriculture and horticulture is now prohibited in countries of the European Union.

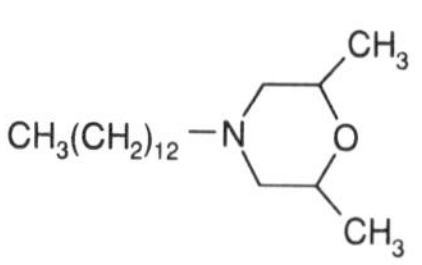

FIGURE 3.12 Tridemorph, a morpholine antifungal. Although it was previously used in agriculture, tridemorph is now banned in many countries because of concern over its toxicity to humans and domestic animals.

Further reading

Anderson, O. S. (1984). Gramicidin channels. *Ann. Rev. Physiol.* **46**, 531.

Baginski M. *et al.* (1997). Molecular properties of amphotericin B membrane channel: a molecular dynamics simulation. *Molec. Pharmacol.* **52**, 560.

Balkis, M. M. (2002). Mechanisms of fungal resistance. *Drugs* **62**, 1025.

Dobler, M. (1981). *Ionophores and Their Structures.* John Wiley & Sons, New York.

Gumila, C. *et al.* (1999). Ionophore–phospholipid interactions in Langmuir films in relation to ionophore selectivity towards *Plasmodium*-infected erythrocytes. *J. Coll. Interface Sci.* **15**, 377.

Hancock, R. E. W. (2001). Cationic peptides: effectors in innate immunity and novel antimicrobials. *Lancet, Infect. Disord.* **1**, 156.

Ingram, L. O. and Buttke, T. M. (1984). Effects of alcohols on micro-organisms. *Adv. Microb. Physiol.* **25**, 254.

Maillard, J.-Y. (2002). Bacterial target sites for biocide action. *J. Appl. Microbiol. Symposium* Suppl. **92**, i.e. 16S.

Odds, F. C., Brown, A. J. P. and Gow, N. A. R. (2003). Antifungal agents: mechanisms of action. *Trends Microbiol.* **11**, 272.

Ridell, F. G. (2002). Structure, conformation and mechanism in the membrane transport of alkali metal ions by ionophoric antibiotics. *Chirality* **14**, 121.

Russell, A. D. (1986). Chlorhexidine, antibacterial action and bacterial resistance. *Infection* **14**, 212.

Schroeder, E. K. *et al.* (2002). Drugs that inhibit mycolic acid biosynthesis in *Mycobacterium tuberculosis. Curr. Pharm. Biotechnol.* **3**, 197 (includes review of triclosan action).

Chapter four

Inhibitors of nucleic acid biosynthesis

Many antimicrobial substances, both synthetic chemicals and natural products, inhibit the biosynthesis of nucleic acids. However, relatively few of these inhibitors are clinically useful as antimicrobial drugs because most do not distinguish between nucleic acid synthesis in the infecting micro-organism and in the host. Many inhibitors of nucleic acid synthesis are therefore too toxic to the host for safe use as antimicrobial agents. However, there are important exceptions which are described in this chapter.

The synthesis of DNA and the various types of RNA is an essential function of dividing and growing cells. Inhibition of DNA synthesis rapidly results in inhibition of cell division. The biosynthesis, recombination and intercellular exchange of extrachromosomal elements of DNA in bacteria are also critical in maintaining the flexible responses of bacteria to changes in the environment (Chapter 8). Inhibition of RNA synthesis is followed by cessation of protein synthesis. The time elapsing between the inhibition of RNA synthesis and the resulting failure of protein biosynthesis can be used to indicate the stability of messenger RNA in intact cells.

Substances that interfere with nucleic acid biosynthesis fall into several categories. The first group includes several effective antibacterial drugs that interfere with the synthesis and metabolism of the 'building blocks' of nucleic acids, i.e. the purine and pyrimidine nucleotides. Interruption of the supply of any of the nucleoside triphosphates required for nucleic acid synthesis blocks further macromolecular synthesis when the normal nucleotide precursor pool is exhausted. Structural analogues of purines and pyrimidines and their respective nucleosides disrupt the supply of correct nucleotides for nucleic acid synthesis and may also directly inhibit nucleic acid polymerization following conversion to the corresponding triphosphates, either by inhibiting DNA- and RNA-dependent polymerase activity or by causing premature chain termination. Few such compounds are useful as antibacterial drugs because of their lack of specificity, but several purine and pyrimidine nucleoside analogues have achieved success as antiviral agents, and a pyrimidine analogue finds application as an antifungal drug. Compounds that interfere with the supply of folic acid also inhibit nucleotide biosynthesis. Interruption of the supply of tetrahydrofolate soon brings nucleotide and nucleic acid synthesis to a halt, and inhibitors of dihydrofolate reductase are useful in antibacterial and antimalarial therapy.

Although DNA-dependent RNA polymerases are common to both prokaryotic and eukaryotic cells, several naturally occurring and semisynthetic antibiotics specifically inhibit the bacterial forms of these enzymes. Another group of inhibitors blocks nucleic acid synthesis by binding to the DNA template. This type of interaction can prevent both DNA replication and transcription into RNA, but, except in the special case of the antibacterial inhibitors of topoisomerases, this is too nonspecific to permit broad therapeutic application.

Finally, several series of compounds, known as topoisomerase inhibitors, block topological changes in bacterial DNA that are essential for the organization and functioning of DNA in cells. These compounds include some of the most valuable antibacterial drugs in current use.

4.1 Compounds affecting the biosynthesis and utilization of nucleotide precursors

4.1.1 The sulfonamide antibacterials

The sulfonamides were the first successful antibacterial drugs. The original observation was made with the dyestuff prontosil rubrum, which is metabolized in the liver to the active drug sulfanilamide. A more effective derivative of sulfanilamide was sulfapyridine, which was in turn superseded by compounds with less toxic side effects. Several of these early compounds are still in use and their structures are shown in Figure 4.1.

Many other sulfonamide antibacterials have been developed since. Most of these are probably no more intrinsically antibacterial than the earlier compounds, although some are much more persistent in the body and can therefore be administered less frequently. The sulfonamides act against a wide range of bacteria, but their main success immediately following their discovery was in the treatment of streptococcal infections and pneumococcal pneumonia. Gradually the sulfonamides were displaced by naturally occurring antibiotics and their derivatives, largely because of the greater antibacterial potency of antibiotics. However, sulfonamides have retained a place in the treatment of certain bacterial and protozoal infections, especially in combination with inhibitors of dihydrofolate reductase. The structural requirements for antibacterial activity in the sulfonamide series are relatively simple. Starting from sulfanilamide, the modifications have generally been variations in substitution on the nitrogen of the sulfonamide group. Substitution on the aromatic amino group causes loss of activity.

Among the many sulfonamides synthesized is dapsone (Figure 4.1) which, although it has no useful action against common infections, has an excellent effect on leprosy and is still a mainstay in the treatment of this disease, in combination with other drugs such as rifampicin to minimize the risk of development of resistant mycobacteria. Dapsone is thought to act by the same biochemical mechanism as the sulfonamides, but the reason for its specificity in leprosy is not known.

A few years after the discovery of the antibacterial activity of the sulfonamides it was shown that some

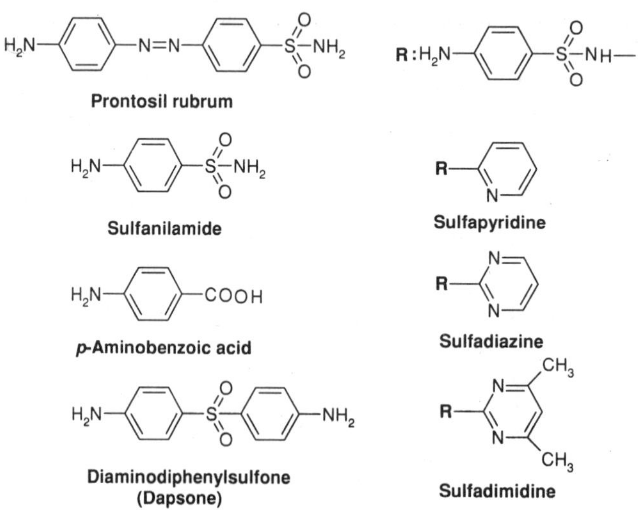

FIGURE 4.1 Examples of sulfonamide antibacterial drugs.

bacteria have a nutritional requirement for *p*-aminobenzoic acid, which is involved in the biosynthesis of folic acid (Figure 4.2). The structural resemblance between *p*-aminobenzoic acid and the sulfonamides underlies the ability of these drugs to antagonize the stimulatory effect of *p*-aminobenzoic acid on the growth of bacterial cells. Later, the structure of folic acid was found to contain a *p*-aminobenzyl group and its biosynthesis was shown to be inhibited by the sulfonamides. The biosynthesis proceeds via the dihydropteridine pyrophosphate derivative shown in Figure 4.2, which then reacts with *p*-aminobenzoic acid with loss of the pyrophosphate group to give dihydropteroic acid. The sulfonamides inhibit the enzyme dihydropteroate synthase (DHPS), which catalyzes this latter reaction in an apparently competitive manner. While DHPS is essential for the *de novo* synthesis of folate in bacteria, yeasts and protozoa, it is absent from mammals, which acquire folate from the diet. DHPS is therefore an excellent target offering high specificity for antimicrobial agents. Almost 60 years after the discovery of competitive antagonism between sulfanilamide and *p*-aminobenzoic acid, the structure of DHPS from *Escherichia coli* and *Staphylococus aureus* was determined. The structure of the enzyme is that of an eight-stranded α/β barrel. The more complex substrate, 7,8-dihydropteridine, binds in a deep cleft in the barrel while sulfanilamide (and presumably *p*-aminobenzoic acid, although this remains to be determined) binds closer to the surface. The structure of DHPS from *Mycobacterium tuberculosis* has also recently been determined in the hope that this may assist in the design of novel antitubercular drugs.

The sulfonamides can also act as alternative substrates, giving rise to reaction products that are analogues of dihydropteroate. However, these products probably do not play a major role in antibacterial action since they inhibit the downstream enzymes only at concentrations higher than those achievable in the cell.

The striking success of the sulfonamides as antibacterials, coupled with the early knowledge of their point of action, led to an extraordinary flurry of chem-

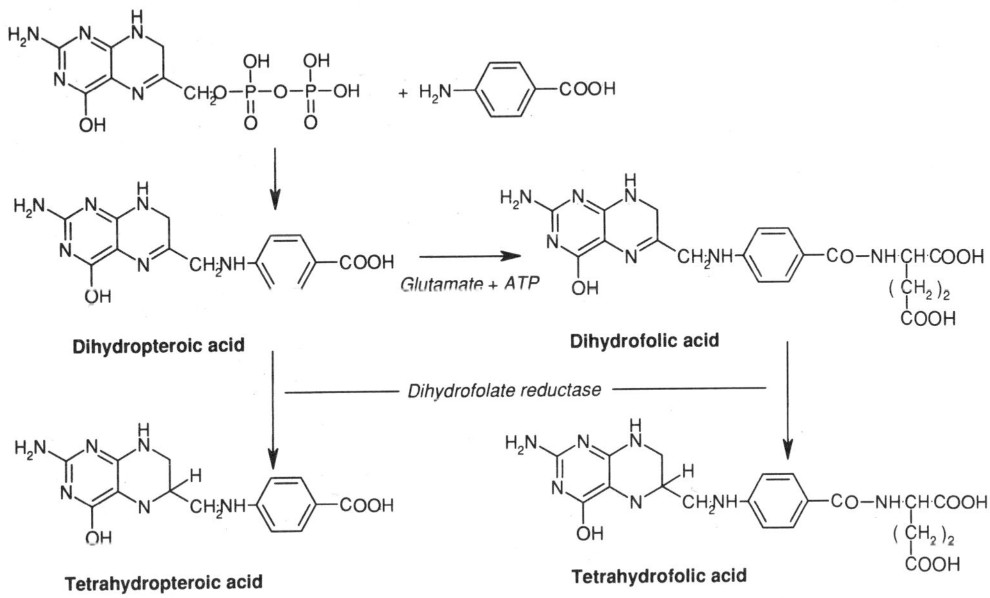

FIGURE 4.2 The final stages of folic acid biosynthesis. The first reaction in the sequence is catalyzed by dihydropteroate synthase, which is competitively inhibited by sulfonamides.

ical research. Every conceivable bacterial growth factor became the model for the synthesis of analogues in the hope of repeating the success of the sulfonamides as antibacterial agents. Unfortunately this early effort was largely fruitless because the apparently simple model provided by the antagonism of *p*-aminobenzoic acid by sulfanilamide was not easily repeated. The sulfonamides owe their effect to a fortunate set of circumstances: (a) As we have seen, DHPS and *p*-aminobenzoate are absent from animal cells, which acquire their folic acid from the diet. (b) The inhibition of bacterial growth is not reversed by folic acid because of its poor diffusion into the cells. In contrast, the sulfonamides, like *p*-aminobenzoic acid, enter bacterial cells freely. (c) Many biosynthetic intermediates carry phosphoric acid groups which tend to prevent their diffusion into bacteria, and potential inhibitors based on analogous structures share the same difficulty of access. This problem was not readily appreciated during the early days of antibacterial drug research.

4.1.2 Inhibitors of dihydrofolate reductase

When the structure of folic acid became known and its relationship to *p*-aminobenzoic acid and the sulfonamides was accepted, a search was made for antagonists among structural analogues of folic acid itself. Many were synthesized but proved to be highly toxic to mammals because folic acid derivatives, in contrast to *p*-aminobenzoic acid, play an important part in the metabolism of animal cells. The toxicity of some of these compounds towards animal cells is actually much greater than towards bacteria since bacterial membranes are almost completely impermeable to them. The cytotoxic action of the antifolic compound methotrexate (Figure 4.3), is useful in the treatment of certain malignancies, rheumatoid arthritis and psoriasis, although great care must be taken to minimize the risk of serious side effects.

Although the direct analogues of folic acid were of no value as antimicrobial agents, other compounds more distantly related to folic acid have considerable importance. The potential of this type of compound was first realized in two drugs developed as antimalarials, proguanil and pyrimethamine (Figure 4.3). Proguanil is a prodrug that is metabolized in the liver to the active agent, cycloguanil (Figure 4.3), which closely resembles pyrimethamine.

The exact point of attack of these so-called antifolic compounds became apparent when the details of folic acid biosynthesis were fully worked out. The

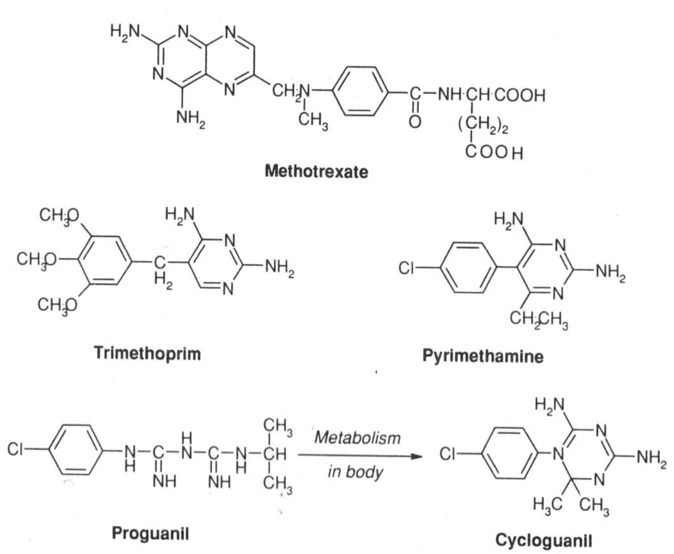

FIGURE 4.3 Drugs that inhibit dihydrofolate reductase. In the case of the antimalarial proguanil, its metabolite, cycloguanil, is the active inhibitor.

step leading to the production of dihydropteroic acid has already been discussed. At this point glutamic acid is added to give dihydrofolic acid, which must be reduced to the tetrahydro state (Figure 4.2) by the enzyme dihydrofolate reductase (DHFR) before it can participate as a cofactor for one-carbon transfer reactions. Cytotoxic analogues of folic acid, such as methotrexate and the antimalarial drugs mentioned earlier, inhibit DHFR. Although most living cells contain DHFR, the enzyme evidently differs in structural details amongst major groups of organisms, and a useful degree of species specificity in the action of inhibitors is possible. For example, pyrimethamine is poorly active against the mammalian and bacterial enzymes, but has an exceptionally strong affinity for the enzyme from the malarial parasite, which accounts for its specific antimalarial action. The antimalarial metabolite of proguanil has analogous specificity for the protozoal DHFR. The structure of DHFR from several species, including *Escherichia coli, Lactobacillus casei,* and *Plasmodium falciparum* in chickens and humans, has been revealed by X-ray analysis. Unlike the DHFR of bacteria and the higher eukaryotes, the enzyme of the malarial parasite is bifunctional, having DHFR and thymidylate synthase domains which are linked by a junctional region. This arrangement may allow the channelling of the product of thymidylate synthase, dihydrofolate, directly to the active site of DHFR for further enzymic processing. Although the DHFR region has amino acid sequence similarities with the DHFR enzymes of other species, there is enough difference to allow preferential inhibition by the antimalarial compounds pyrimethamine and cycloguanil. Full details of the crystallographic analysis of the interaction between pyrimethamine and the DHFR of *Plasmodium falciparum* are available in a reference listed in 'Further reading.' Pyrimethamine is used in combination with the sulfa drug sulfadoxine in the treatment of malaria whereas proguanil is combined with chloroquine (Chapter 6).

A highly selective compound against the bacterial DHFR is the pyrimidine derivative trimethoprim (Figure 4.3). Reduction of the activity of bacterial dihydrofolate reductase by 50% requires a trimethoprim concentration of 0.01 μM, whereas the same inhibition of the human enzyme requires 300 μM. A recent comparison of a bacterial DHFR (from *Mycobacterium tuberculosis*) with its human counterpart revealed that the bacterial enzyme contains 159 amino acid residues compared with 187 in the human protein. Even though the human enzyme is significantly longer, the two proteins exhibit the same general folding of their three-dimensional structures. However, there is overall only 26% sequence identity, and the two enzymes have many local regions of difference in their three-dimensional structures, including around the active sites. X-Ray analysis shows that trimethoprim binds to the bacterial enzyme at the active site, with its pyrimidine ring held in a deep cleft by hydrogen bonds and van der Waals interactions with neighbouring amino acids. The trimethoxybenzyl side chain of the drug extends out towards the entrance of the cleft and forms van der Waals interactions with the amino acids of two separate helical regions of the enzyme. Presumably because trimethoprim binds so weakly with the human DHFR, it has not been possible to study the interaction by X-ray analysis.

Trimethoprim is effective on its own as an antibacterial drug but is more generally used as a combination (cotrimoxazole) with the sulfonamide derivative sulfamethoxazole. The combination is claimed to have a wider field of antibacterial activity than either compound alone and is prescribed as a broad-spectrum alternative to ampicillin. Because the sulfonamide and trimethoprim block the folic acid biosynthetic pathway at different points, the twofold inhibition is especially effective in depriving bacteria of tetrahydrofolate. The principle of twofold blockade is also exploited in the antimalarial combination of pyrimethamine with sulfadoxine mentioned earlier.

The reduction in tetrahydrofolate levels in bacteria, fungi and protozoa caused by sulfonamides and the dihydrofolate reductase inhibitors has widespread effects on cells, although the antifolate approach has not yet been successfully exploited in the treatment of fungal infections. Tetrahydrofolate is required as a one-carbon unit donor in the biosynthesis of methionine, glycine and the formyl group of fMet-tRNA. Tetrahydrofolate deprivation therefore depresses protein synthesis. The major effects, however, are on the biosynthesis of purines and pyrimidines, which involve one-carbon transfer reactions at several stages. The synthesis of thymine is particularly sensitive to inhibitors of dihydrofolate reductase because of the requirement for tetrahydrofolate in the transformation of dUMP to dTMP (Figure 4.4). When cultures of bacteria are grown in media containing amino acids and inosine, antagonists of folic acid synthesis cause the phenomenon known as thymineless death, which can be prevented by the addition of excess thymine or thymidine.

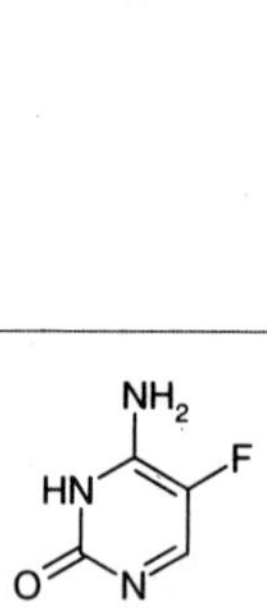

FIGURE 4.5 The antifungal drug 5-fluorocytosine.

4.2 Nucleoside analogues

4.2.1 5-Fluorocytosine (flucytosine, 5-FC)

This pyrimidine analogue (Figure 4.5) was originally synthesized as an anticancer agent but is now used mainly to treat certain serious pathogenic yeast infections, including cryptococcosis and candidiasis. Except in the latter infection, 5-FC is usually administered in combination with amphotericin B (Figure 3.8). It is not itself active and must be metabolized to compounds that are the effective inhibitors. The uptake of 5-FC into fungal cells is facilitated by the transporter protein cytosine permease. Subsequently, the compound is converted to 5-fluorouracil by cytosine deaminase. Fortunately this enzyme is absent from human cells since 5-fluorouracil is highly toxic to all dividing mammalian cells. 5-Fluorouracil enters a complex network of fungal nucleotide metabolism. An important end product is 5-fluorodeoxyuridine-5′-monophosphate, which inhibits thymidylate synthase and therefore DNA synthesis. The other major route of metabolism that contributes to the antifungal activity of 5-FC is via the conversion to 5-fluorouridine-5′-monophosphate, catalyzed by UMP phosphoribosyl transferase, leading to the incorporation of 5-fluorouridine-5′-triphosphate into RNA. The antifungal action of 5-FC therefore results from a combination of the inhibition of DNA synthesis and the generation of aberrant RNA transcripts. Filamentous fungi generally lack the suite of enzymes necessary for the uptake and metablism of 5-FC and are therefore unaffected by the drug.

During therapy with 5-FC, careful monitoring of blood levels is important to ensure that concentrations toxic to the kidneys and bone marrow are not achieved, while levels are still high enough to suppress the infection and minimize the risk of the emergence of 5-FC-resistant mutants.

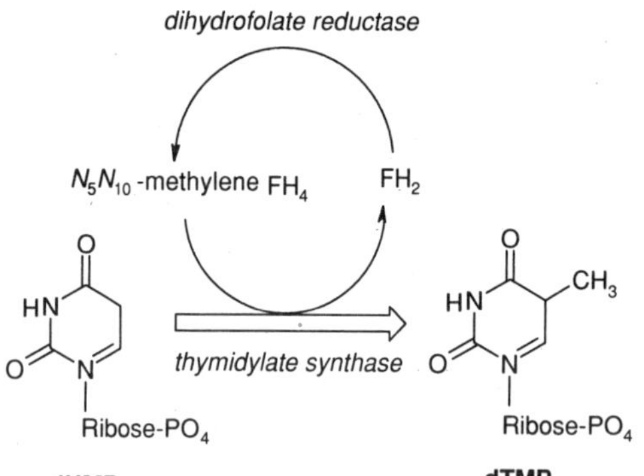

FIGURE 4.4 A major consequence of the inhibition of dihydrofolate reductase is the suppression of thymine biosynthesis. Bacteria cultured with trimethoprim undergo 'thymineless death' when not supplemented with thymine or thymidine.

4.2.2 Antiviral nucleoside analogues

Several nucleoside analogues have achieved considerable success as antiviral drugs. Like 5-FC, all these compounds are prodrugs that are metabolized by host or viral enzymes to the active inhibitors.

Ribavirin

This nucleoside analogue (1-β-D-ribofuranosyl-1,2,4-triazole-3-carboxamide; Figure 4.6) is an antiviral drug effective against certain DNA and RNA viruses. It has long been used to treat serious lung infections in young children caused by the respiratory syncytial virus. In combination with interferon-α, ribavirin also has an important application in the treatment of hepatitis C. Ribavirin has direct actions against viruses with

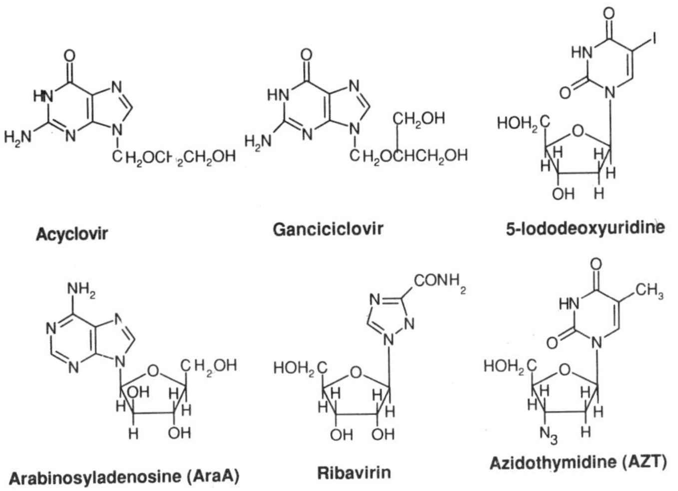

FIGURE 4.6 Examples of nucleoside analogues as antiviral drugs.

multiple sites of interference in virus replication and indirect effects in modulating the host's immune defences against viral infection.

Direct antiviral action. Ribavirin is successively phosphorylated by enzymes of the host cells to the corresponding mono-, di- and triphosphates. The triphosphate is recognized as a substrate mimic for viral polymerases, resulting in misincorporation into the viral nucleic acids. Among other effects, misincorporation causes premature termination of the RNA primer chain essential for DNA synthesis. Misincorporation of ribavirin nucleotides into viral RNA may also result in viral mutagenesis. There is some evidence for this effect in the polio RNA virus, which markedly reduces the infectivity of the virus. While ribavirin is not used to treat infection by the polio virus, the mutagenic action of the drug could be relevant to its clinical efficacy against hepatitis C, although there is no direct evidence for this.

In some viruses the viral messenger RNA is modified at the 5′ end by a 'cap' which protects the mRNA from degradation by host nucleases. Capping involves the addition of GTP to the 5′-hydroxyl group of RNA to form a 5′- to 5′ triphosphate link catalyzed by guanyl transferase. Subsequent methylation of guanine at the 7 position completes the capping process. Ribavirin triphosphate may block the capping process in some viruses, e.g. vaccinia, by competing with GTP at the active site of the guanyl transferase.

The initial conversion of ribavirin to the monophosphate by host cell adenosine kinase provides another opportunity for interference with viral replication. Ribavirin monophosphate is a weak competitive inhibitor of another host enzyme, inosine 5′-monophosphate dehydrogenase (IMPDH), which catalyzes a rate-limiting reaction in the biosynthesis of guanine nucleotides:

$$\begin{matrix}\text{inosine} \\ \text{5}'\text{-monophosphate}\end{matrix} + NAD^+ \leftrightarrow \begin{matrix}\text{guanosine} \\ \text{5}'\text{-monophosphate}\end{matrix} + NADH^+.$$

Limitation of the intracellular availability of guanine nucleotides by even partial inhibition of the above reaction is likely to exert multiple negative effects on viral replication, including the capping processs and nucleic acid polymerization.

Indirect antiviral action. There is evidence that ribavirin enhances the host's immune defences against the progression of viral infection. A detailed discussion of this aspect of ribavirin action is outside the scope of this book, but in summary, the drug promotes T-lymphocyte immunity by inducing the production of several antiviral type I cytokines, including interferon-γ, tumor necrosis factor-α and interleukin-2. At the same time, ribavirin treatment suppresses the proviral type 2 cytokines, such as interleukin-4 and interleukin-10. The inhibition of lymphocyte IMPDH and the consequent depression in cellular GTP content is thought to underlie these complex effects.

The value of ribavirin as an antiviral drug has slowly emerged over many years and its site of action has often been the subject of controversy. We can now begin to see that the clinical benefits of the drug owe much to a combination of the multiplicity of its effects on viral infection, both direct and indirect.

Acyclovir and ganciclovir

These structurally similar compounds (Figure 4.6) are analogues of guanosine but without the cyclic ribose group. Despite their similarity, the two drugs have different clinical applications. Acyclovir is used to treat herpes simplex and varicella-zoster infections. While ganciclovir is also active against herpes viruses, its clinical application is limited to the treatment of cytomegalovirus infections, which are particularly troublesome in AIDS patients.

Herpes viruses code for a virus-specific form of thymidine kinase which converts acyclovir and ganciclovir to their monophosphate derivatives. It is significant that the thymidine kinase of the host cells has a much lower substrate affinity for these compounds, so that uninfected cells do not generate the phosphorylated derivatives. The drug monophosphates in virus-infected cells are then successively phosphorylated by host cell kinases to the triphosphate level. The triphosphates of both drugs are good substrates for the DNA polymerases encoded by herpes viruses but are only poorly recognized by the host polymerases. The resulting preferential incorporation of the drug triphosphates into viral DNA causes premature chain termination because the antiviral nucleotides lack a 3′-OH group on their sugar residues, thus preventing formation of the 3′-5′-phosphodiester bonds necessary for chain extension.

The action of ganciclovir against cytomegalovirus is rather different. In cytomegalovirus-infected cells the drug is again first converted to the monophosphate, though not by thymidine kinase, which the cytomegalovirus lacks. Instead, it is believed that another viral kinase, encoded by the *UL97* gene, may be responsible for the first-stage phosphorylation. The subsequent stages in the metabolism of ganciclovir monophosphate and its antiviral action resemble those in herpes-infected cells.

Vidarabine

This drug (9-β-D-arabinosyladenine, AraA; Figure 4.6) has a fairly broad antiviral spectrum, including activity against herpes viruses, cytomegalovirus and the Epstein-Barr virus. Its triphosphate metabolite inhibits viral DNA polymerase, and viral ribonucleotide reductase and is also incorporated into viral DNA. Because these effects occur at concentrations of vidarabine below those needed to inhibit host DNA synthesis, the compound can be used topically against herpes infections of the eye and brain.

5-Iododeoxyuridine

This is another nucleoside (Figure 4.6) which has been used against herpes infections, especially of the cornea, but is less specifically antiviral than the compounds described earlier. The similar van der Waals radii of the iodine atom (0.215 nm) and the methyl group (0.2 nm) of thymidine enable the drug to replace thymidine in DNA with considerable efficiency since its triphosphate is readily accepted as a substrate by DNA polymerase. The incorporation of 5-iododeoxyuridine into viral DNA leads to errors of replication and transcription and the eventual termination of viral proliferation. Unfortunately the compound is also incorporated into host DNA and the resultant toxicity limits its use to topical application to the eye and skin.

4.3 Inhibitors of the reverse transcriptase of the human immunodeficiency virus

The search for effective treatments against the virus that causes AIDS has been, and continues to be, one of the greatest therapeutic challenges of modern medicine. The progress of the infection and its associated pathology are insidious and irreversible damage to the immune system may occur if the disease is not diagnosed sufficiently early. Nevertheless, considerable progress has been achieved in developing drugs that limit viral proliferation and confer significant clinical benefits. Currently, 16 drugs are approved for the treatment of HIV infection. Ten of these are inhibitors of the characteristic reverse transcriptase (RT) of HIV which transcribes the single-stranded RNA genome into a single-standed DNA and subsequently synthesizes a complementary strand of DNA. The resulting double helical DNA is then capable of integration into the chromosomes of the host cell. The other six drugs inhibit the viral protease and are discussed in Chapter 6. The inhibitors of reverse transcriptase include six nucleoside analogues, one nucleotide analogue and three non-nucleoside compounds (Figures 4.6, 4.7, 4.8). The rapid acquisition by HIV of resistance to any single drug (see Chapter 7), and the consequent and ominous likelihood of therapeutic failure, now mandates the exclusive use of the multidrug regime known as highly active antiretroviral therapy or HAART, combining inhibitors of both RT and the viral protease. Nevertheless, the search for new and improved anti-HIV drugs continues apace and many new inhibitors of RT and other agents are currently under investigation.

4.3.1 Azidothymidine

This drug (3′-azido-3′-deoxythymidine; AZT, Figure 4.6) was the first effective anti-HIV drug to be discovered. The compound was originally developed as a potential anticancer drug but was then found to be effective against the replication of HIV in AIDS patients. The compound is first efficiently converted by the host

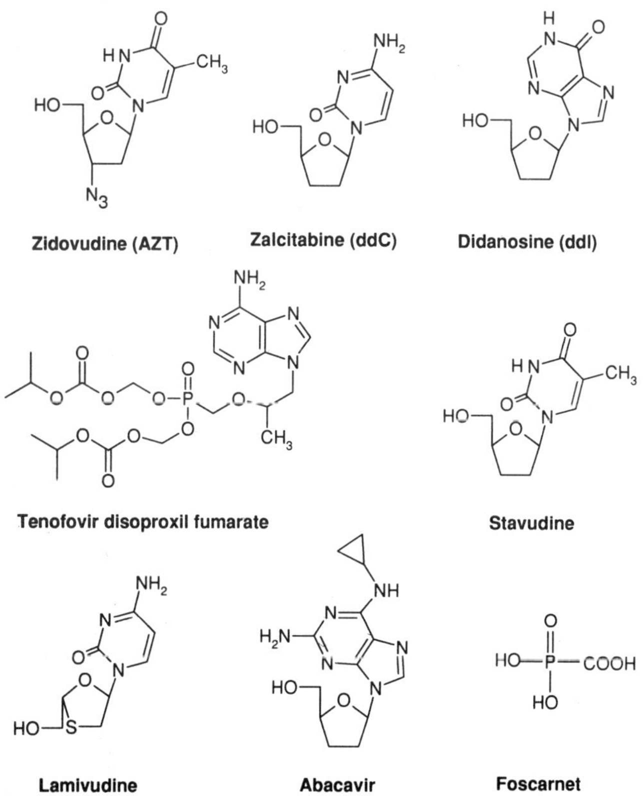

FIGURE 4.7 Inhibitors of the reverse transcriptase of HIV used in the treatment of HIV infection and AIDS. These agents act at the active center of the enzyme.

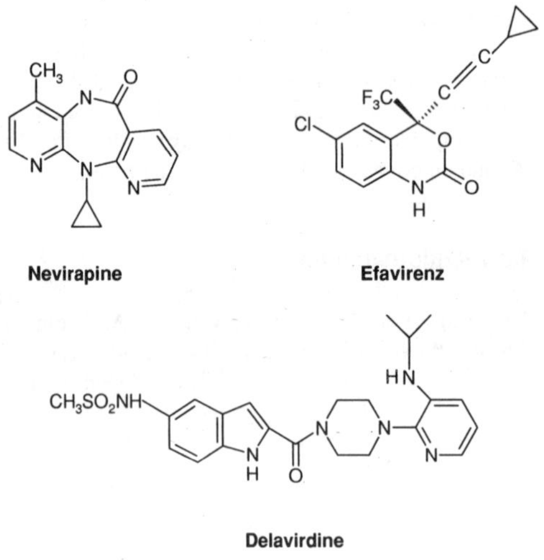

FIGURE 4.8 Non-nucleoside inhibitors of the reverse transcriptase of HIV used in the treatment of HIV infection and AIDS. These agents bind in a pocket sited away from the active center of the enzyme.

cell thymidine kinase to the monophosphate derivative, AZT-MP. The subsequent phosphorylations to the di- and triphosphate derivatives are also carried out by a host cell enzyme, thymidylate kinase. AZT-MP is, however, a relatively poor substrate for this enzyme, in part because the larger size of the 3′-azido group of AZT-MP compared with the 3′-OH group of thymidine monophosphate hinders the interaction of AZT-MP with the active site of thymidylate kinase. HIV RT has three distinct enzymic activities:

1. An RNA-dependent DNA polymerase,
2. A ribonuclease (H),
3. A DNA-dependent DNA polymerase.

Respectively, these activities:

1. Copy the plus-strand RNA of the virus to produce a minus-strand DNA,
2. Remove the RNA template,
3. Synthesize the plus-strand of DNA using the minus-strand DNA as a template.

Reverse transcription is a dimeric enzyme with subunits of 65 kDa (labeled p66) and 51 kDa (p51). Crystallographic analysis of the enzyme, which was achieved only after overcoming considerable difficulties in crystallizing the protein, reveals that both subunits contain domains referred to as palm, fingers, thumbs and connection. An additional C-terminal domain in the p66 subunit provides the RNAase activity whereas the polymerase function is located in the palm domain of p66. The p66 palm domain, together with the p66 fingers and thumbs, form a cleft that binds the template–primer complex in proximity to the polymerase-active site. The polymerase is believed to carry out both RNA- and DNA-dependent processes. The triphosphate of AZT effectively competes with the endogenous thymidine triphosphate at the polymerase-active site, resulting in the incorporation of AZT-monophosphate into the viral DNA. Further extension of the DNA chain is then blocked because the absence of a hydroxyl group in the 3′ position of the ribose moiety of AZT precludes the formation of the next 3′-5′-phosphodiester bond in DNA .

Because there is no reverse transcriptase in uninfected host cells, it might appear that AZT would be specifically active against the virus infection. Unfortunately, the clinical use of AZT is beset with problems. First, virus replication is only reduced to about 10% of the normal rate, largely because of the relatively inefficient conversion of AZT-MP to the antiviral triphosphate. The incomplete inhibition of viral replication facilitates the rapid emergence of AZT-resistant mutants. There is also a major problem of bone marrow toxicity, which is probably due to the interference of AZT phosphates with the pyrimidine metabolism of host cells and with host cell DNA synthesis.

The other anti-HIV drugs shown in Figure 4.7, with the exception of foscarnet, are all considered to act in a manner similar to that of AZT. They are progressively phosphorylated to the triphosphates and then compete with the corresponding endogenous nucleotides at the polymerase-active site of RT for incorporation into viral DNA. The absence of the essential hydroxyl group at the ribose 3′-position in all the compounds again results in premature termination of the chain. Tenofovir disoproxil fumarate (TDF) is an unusual nucleotide prodrug; the fumarate moiety facilitates diffusion of the drug across cell membranes. Cellular esterases then release the tenofovir monophosphate entity, which is converted to the active triphosphate. Several of the RT inhibitors damage the host mitochondria, apparently owing to an affinity of the triphosphate derivatives of the drugs for DNA polymerase γ of mitochondria. Tenofovir triphosphate exhibits very low affinity for this enzyme, at least *in vitro*, indicating that the drug may be less toxic *in vivo* than other RT inhibitors. The inhibition of RT by the chemically very simple drug foscarnet results from a structural analogy with pyrophosphate which enables the drug to compete with the endogenous nucleoside triphosphates at the pyrophosphate binding site on the RT, thus blocking the growth of nucleic acid chains. Foscarnet is a broad-spectrum antiviral drug with useful activity against cyclomegalovirus infection of the retina and acyclovir-resistant herpes, in addition to its anti-HIV activity.

4.3.2 Non-nucleoside inhibitors of HIV reverse transcriptase (NNRTIs)

The discovery of this class of compounds heralded a new phase in the exploitation of RT inhibition in anti-HIV therapy. Not only are NNRTIs novel chemicals (Figure 4.8) quite different from the nucleoside and nucleotide inhibitors (NRTIs), they also provide an escape from the specific problem of resistance caused by mutations at, or near to, the active center of the polymerase. Unlike the NRTIs, the NNRTIs are noncompetitive inhibitors that interact with the enzyme at a site distant from the active center. The first NNRTI to be developed was nevirapine (Figure 4.8). X-Ray crystallography shows that this compound binds in a pocket located between two β-sheets of the palm of the p66 subunit some 10 Å away from and just under floor of the polymerase active center. The internal face of the pocket is lined with hydrophobic amino acids, including leucine, valine, tryptophan and tyrosine. The NNRTIs are chemically diverse and yet appear to share a generally common mode of binding that is probably largely facilitated by the hydrophobic nature of the binding pocket and flexibility in the protein chains that form it. The interaction process between enzyme and inhibitors is probably accompanied by conformational rearrangements in the drug molecules themselves as they move from the solution phase into the binding pocket. Inhibition of RT activity results from adverse changes in the shape of the active center which derive from the conformational rearrangements involved in the binding of the NNRTI in the binding pocket. This explains the noncompetitive mode of inhibition exerted by NNRTIs.

Valuable as the introduction of NNRTIs has been in the treatment of HIV infection, the drugs have brought their own particular problems. Resistance resulting from new mutations affecting RT away from the active center soon emerged (Chapter 7), and the drugs cause a range of adverse side effects. Nevertheless, the NNRTIs are well established as an essential component of the current HAART regimes.

4.4 Antibacterial inhibitors of topoisomerases

The replication of double-stranded (duplex) DNA in prokaryotes and eukaryotes is a remarkably complex process involving an elaborate suite of proteins and enzymes. The whole process, from bacteria to mammals, shares many common features which indicate a highly conserved nature throughout a long evolutionary history. Even so, it is increasingly clear from genomic studies that there are significant differences among the corresponding enzymes of divergent species. It is therefore rather surprising that relatively few leads have emerged either from natural sources or from synthetic chemistry which target the enzymes of DNA replication that provide specific antimicrobial drugs. Although the increasing research attention being paid to species differences among the replicative enzymes may eventually improve this situation, the only successful drugs so far are the species-specific inhibitors of DNA topoisomerases.

DNA cannot exist in bacterial cells as an extended double-helical molecule. The length of bacterial DNA is about 1300 μm and typically the cell into which it fits is about 1 μm in diameter. Clearly there must be a high degree of ordered quaternary structure in the DNA to accommodate it within cells. This is achieved by negatively supercoiling the DNA; i.e. the supercoiling is left-handed, in contrast to the right-handed winding of the double helix. Special enzymes and proteins induce torsional stresses in the molecule. In this way, the enzymes alter the three-dimensional shape of DNA while maintaining its primary structure and the genetic information encoded in it. These enzymes are also essential for DNA replication and transcription. Initiation of DNA replication can only start if the DNA is negatively supercoiled. This is because negatively supercoiled molecules of the nucleic acid are easier to 'melt' locally than relaxed DNA in order to generate a single-stranded template region. This enables the protein referred to as DnaA to promote the interaction of DNA with the site of origin of replication. Second, when a circular supercoiled DNA molecule is replicated, the two daughter molecules become interlocked, or 'catenated', and without a means of removing the supercoils, the separation of the progeny (decatenation) is impossible.

The enzymes that facilitate these topological changes are known as topoisomerases, four types of which have been identified in bacteria. The type II topoisomerase, or DNA gyrase, is the only one of the four that can introduce negative supercoils. It consists of two 97 kDa GyrA chains and two 90 kDa GyrB chains and is a major target for several classes of antibacterial drugs. The corresponding enzyme in mammalian cells is not significantly targeted by the antibacterial inhibitors of DNA gyrase. The decatenation of interlocked daughter DNA molecules in bacteria is catalyzed by another enzyme, topoisomerase IV.

The supercoiling reaction begins with a segment of DNA, approximately 120 base pairs (bp) in length, wrapping itself around the tetrameric complex of the gyrase. The GyrB subunit allows passage of the DNA segment into the interior of the enzyme using enzyme-bound ATP as an energy source. Both strands of DNA are then cleaved by GyrA. The 5′-phosphate terminal of each cleaved strand is covalently bonded to the hydroxyl group of tyrosine-122 in each of the GyrA subunits. This link is essential to prevent the free rotation of the cut DNA strands. The enzyme next permits a segment of double-stranded DNA to pass through the gap of the broken sequence. The ends of the cleaved strands are then brought together and resealed by the ligase activity of the enzyme. Finally, the GyrB subunit catalyzes the hydrolysis of the bound ATP, permitting the release of the processed DNA segment.

4.4.1 Quinolones

These compounds compose one of the most important groups of wholly synthetic antibacterial drugs in current medical use. Nalidixic acid and oxolinic acid (Figure 4.9) are the so-called first-generation quinolones, whose spectrum of antibacterial action is confined to Gram-negative bacteria. The introduction of a fluorine atom at position C-6 in the second-generation compound ciprofloxacin (Figure 4.9) resulted in a marked increase in potency and extended the antibacterial spectrum to important Gram-positive pathogens. Substitution with basic substituents to counteract the acidity of the carboxyl group, as in gemifloxacin (Figure 4.9), further increased the potency against streptococci and staphylococci. Current interest in the quinolones is largely directed at developing compounds with activity against bacteria resistant to the earlier quinolones.

The antibacterial activity of the quinolones is primarily due to inhibition of DNA gyrase. When the isolated enzyme is incubated with DNA and a quinolone, the supercoiling reaction is arrested at the point where the cut ends of the DNA strands are covalently linked to the hydroxyl groups of the tyrosine-122 residues of GyrA. The re-ligation of the broken strands is blocked and the supercoiling reaction can be said to have been frozen midway. This results in the accumulation of double-stranded nicks in the bacterial genome and may also prevent the essential movements of DNA and

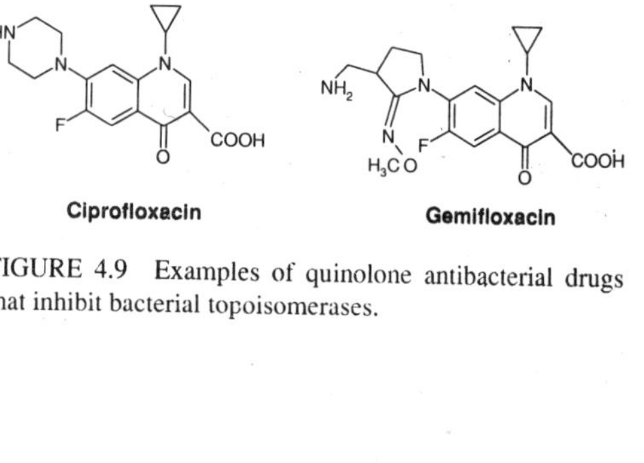

FIGURE 4.9 Examples of quinolone antibacterial drugs that inhibit bacterial topoisomerases.

RNA polymerases along the DNA template. The bactericidal action of the quinolones probably arises from a combination of these effects.

Much attention has been devoted to unravelling the complexities of the enzymic mechanisms of DNA gyrase—the binding sites and modes of inhibition of various antibacterial inhibitors of the enzyme. The study of mutations affecting the activity of DNA gyrase and its resistance to inhibitors, combined with data from X-ray analysis of crystallizable domains of the two subunits, has provided at least partial descriptions of the catalytic process and how it is blocked by antibacterial drugs. The following discussion concentrates largely on the topic of drug action; the reader should consult references listed under 'Further reading' for detailed descriptions of gyrase structure and function.

The primary site of quinolone action is assigned to the GyrA subunit because the most common mutations that confer resistance to these drugs are found in a region of GyrA referred to as the quinolone resistance-determining region or QRDR. For example, replacement of serine-83, which lies close to the active site where DNA is bound by tryptophan, gives about a 20-fold increase in resistance to many different quinolones. Quinolones bind strongly to gyrase complexed with DNA but only weakly to either the enzyme or DNA alone, suggesting that the drugs interact simultaneously with both the protein and nucleic acid. In fact, quinolones are believed to bind to a pocket consisting of the QRDR and a segment of DNA which is also bound to it. A recent refinement of this model for the interaction of ciprofloxacin proposes that the planar character of the bicyclic moiety of the drug enables two molecules of the compound to stack, or intercalate (see later discussion) between the bases of the DNA bound at the active site. Alternatively, the drug might actually displace a cytosine residue from the double helix opposite the bond cleaved by the enzyme (Figure 4.10). The drug is now positioned to interact with several amino acid residues of GyrA, including serine-83 and aspartate-87. Some mutations in GyrB which can also confer quinolone resistance suggest that a limited region of this subunit may contribute to the structure of the binding pocket. Crystal structures of a 59-kDa fragment of the N-terminal domain of GyrA and a 43-kDa fragment of the N-terminal domain of GyrB are

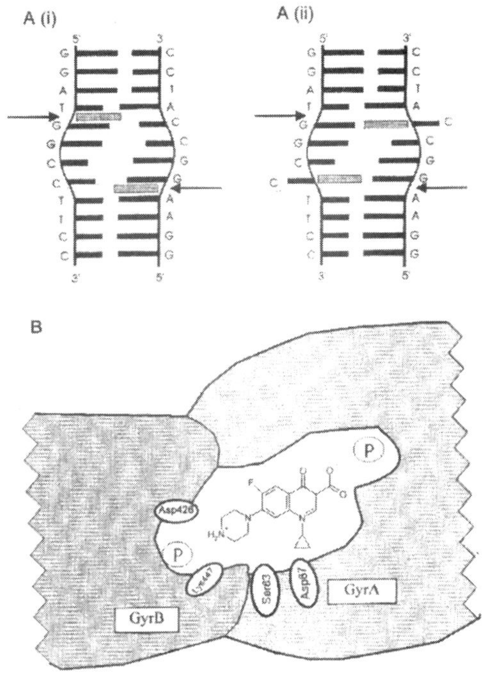

FIGURE 4.10 A possible model of the quinolone binding pocket. A(i): The drug (gray rectangle) intercalated into the DNA double helix. A(ii): In this alternative model of intercalation, the drug has displaced a cytosine residue which is 'flipped out' of the helix. (B): The binding pocket. The GyrB subunit is positioned close to the quinolone resistance-determining region of the GyrA subunit, allowing aspartate-426 and lysine-447 to interact with the bound quinolone molecule. The DNA axis is perpendicular to the plane of the diagram, and the approximate position of the sugar-phosphate backbone is shown by the encircled Ps.The drug interacts with the distorted region of DNA at the active site, and the bottom edge of the quinolone can also interact with serine-83 and aspartate-87 of GyrA. [Taken with permission from J. Heddle and A. Maxwell *Antimicrobial Agents and Chemotherapy* **46**, 1805–1815 (2002).]

available. However, a full crystallographic analysis of a DNA-gyrase-DNA-quinolone complex is still awaited. Only then will the details of the quinolone binding site be finally revealed.

Although the type II topoisomerase is believed to be the major target for quinolones, the type IV enzyme is also inhibited by some fluoroquinolones, although to a lesser extent than the type II enzyme. Indeed in *Staphylococcus aureus* the principal target for quinolones is believed to be topoisomerase IV. In *Streptococcus pneumoniae* the inhibition of both DNA gyrase and topoisomerase IV contribute to the antibacterial action of quinolones. The amount of the contributions varies with the individual quinolone.

Type IV topoisomerase does not induce negative supercoiling of DNA, but plays an important role in the partitioning of DNA during cell division. Its principal catalytic function is to ensure the decatenation of DNA that is essential for the correct separation of DNA into the daughter cells. Despite the distinct functions of the two enzymes, there is extensive amino acid sequence homology between them, which may underlie their susceptibility to inhibition by quinolones. The type IV enzyme is a heterotetrameric protein composed of two subunits, ParC and ParE. ParC and ParE are homologous to GyrA and GyrB with 36 and 40% sequence identity, respectively (in *Escherichia coli*). Currently there are no details of the interaction between quinolones and type IV topoisomerase, but because of its homology with DNA gyrase, the drugs may bind to both enzymes in a generally similar way. Quinolones stimulate DNA cleavage by the type IV enzyme while at the same time inhibiting the religation action.

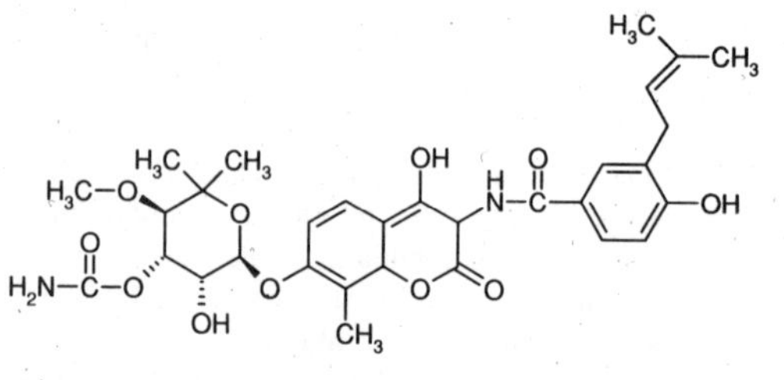

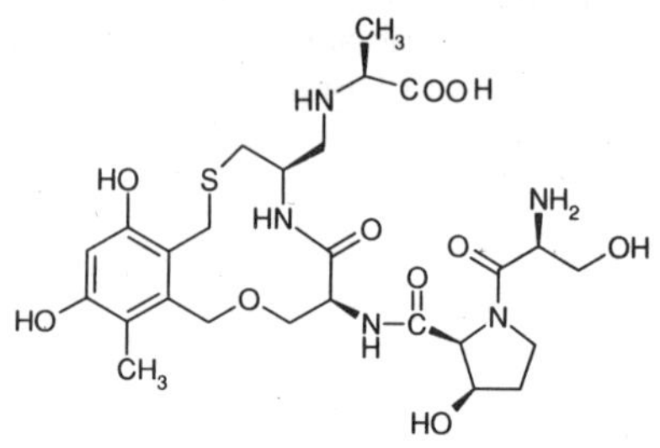

FIGURE 4.11 Naturally occurring inhibitors of the function of the β subunit of bacterial DNA gyrase.

4.4.2 Inhibitors of the GyrB subunit of DNA gyrase: coumarins and cyclothialidine

The disturbing frequency of mutations to quinolone resistance that arise in the GyrA subunit drives considerable interest in the GyrB subunit as an attractive alternative target for drug action. The coumarins, exemplified by novobiocin (Figure 4.11) and cyclothialidine (Figure 4.11), are members of structurally distinct families of naturally occurring inhibitors of DNA negative supercoiling; their target is the ATP hydrolyzing activity of GyrB. Studies on the kinetics of inhibition of DNA gyrase by both types of compound indicate potent competition with ATP, with inhibitor constant (K_i) values in the range of 10^{-7} to 10^{-9} M. However, examination of the ATPase activity of isolated GyrB subunits indicated that it did not follow Michaelis-Menten kinetics, thus raising doubts about the concept of competitive inhibition. These doubts were supported by the absence of any significant structural similarity between the inhibitors and ATP. Furthermore, point mutations in GyrB that cause resistance to the coumarins were found to lie at the periphery of the ATP binding site. Fortunately, the uncertainty was eventually resolved by X-ray crystallographic data on the complexes of the GyrB subunit with novobiocin, cyclothialidine and adenylyl-β-γ-imidodiphosphate (an analogue of ATP). Although the three ligands are structurally distinct and bind to the enzyme in very different ways, the X-ray analysis shows that there is some overlap of their binding sites. The concept of competitive inhibition of ATP binding to the B subunit by both the coumarins and cyclothialidine may therefore be correct. As would be expected, the interactions of novobiocin and cyclothialidine with the enzyme are complex and the reader is referred to the research paper listed under 'Further reading' for detailed information. To summarize, in the case of novobiocin, the noviose sugar unit contributes to extensive hydrogen bonding with key amino acids of the enzyme and water molecules associated with the protein. Hydrophobic interactions involving both the sugar and coumarin elements add to the stability of the complex. The novobiocin molecule does not lie flat in the complex but is bent around on itself in a manner that has been likened to a staple. This is due to the conformation of the peptide bond between the coumarin ring and the hydroxybenzoate group being in the *cis* conformation rather than in the expected *trans* conformation. The binding contribution of the hydroxybenzoate group is apparently the least significant because a chemically modified derivative of novobiocin lacking this group still inhibits both DNA supercoiling and the ATPase activity. The antibacterial activity of this compound is, however, much reduced and it is possible that the 3′-isopentenyl-4′-hydroxybenzoate group facilitates the entry of novobiocin into bacterial cells.

The binding of cyclothialidine to the GyrB subunit is stabilized largely by the formation of water-mediated hydrogen bonds between the hydroxyl groups of the substituted resorcinol group of the inhibitor and key amino acids that line the binding site which overlaps with that for ATP. The clinical history of novobiocin and other coumarins has been disappointing, owing to several factors. The coumarins are poorly absorbed in oral doses and their antibacterial activity is limited mainly to Gram-positive organisms, probably because of limited penetration into Gram-negative cells. Drug-resistant mutants emerge readily from initially drug-sensitive populations of Gram-positive bacteria. Finally, although there is no functional equivalent of DNA gyrase in mammalian cells, mammalian topoisomerase II, which shares some sequence homology with gyrase, is susceptible to inhibition by novobiocin. In contrast, cyclothialidine is much more specific for the bacterial enzyme, offering hope that the design of specific inhibitors of GyrB function may lead to novel and effective antibacterial drugs. However, cyclothialidine penetrates bacterial membranes very poorly and is only weakly antibacterial in consequence. Recent chemical modification of cylcothialidine, however, has produced several compounds with much improved antibacterial activity *in vitro* and *in vivo*.

4.5 Inhibitors of DNA-dependent RNA polymerase

The transcription of RNA from DNA is, of course, common to both prokaryotic and eukaryotic organisms and involves enzymes known as DNA-dependent RNA polymerases. It was therefore somewhat surprising to discover that an important group of antibacterial antibiotics, the rifamycins isolated from *Streptomyces*, show remarkable specificity for the inhibition of bacterial DNA-dependent RNA polymerase. One of these compounds, rifampicin, or rifampin (Figure 4.12), is a potent antibiotic and a mainstay of treatment for tuberculosis in combination with other drugs. Rifampicin is active against many Gram-positive bacteria, but is less

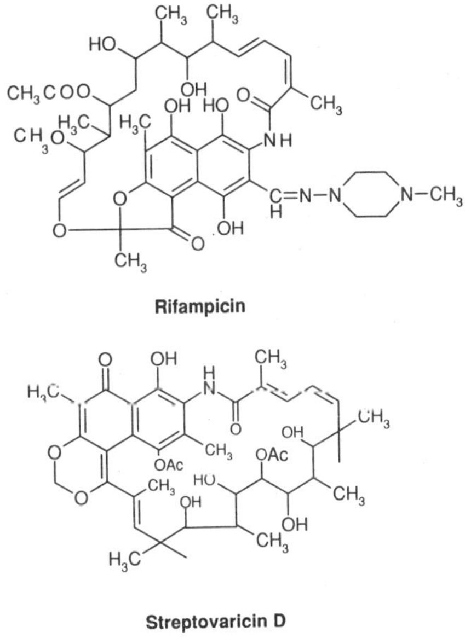

FIGURE 4.12 Two antibiotics that selectively inhibit DNA-dependent RNA synthesis in bacteria. Rifampicin is a semisynthetic member of the rifamycin group. Streptovaricin D is related in structure to the rifamycins; jointly the rifamycins and streptovaricins are known as ansamycins.

effective against Gram-negatives because of limited access to the target enzyme in these organisms. Chemically the rifamycins are closely related to the streptovaricins (Figure 4.12). The two groups of antibiotics appear to have a similar mode of action, but unlike the rifamycins, the streptovaricins are not in medical use. Both groups strongly inhibit RNA synthesis in sensitive bacteria and in cell-free extracts by binding to and inhibiting DNA-dependent RNA polymerase. The drugs neither bind to nor inhibit the corresponding mammalian enzyme.

The RNA polymerase isolated from *Escherichia coli* is a large (450 kDa), complex enzyme consisting of four kinds of subunits: α, β, β' and σ. The complete or holoenzyme has the composition ($\alpha_2\beta\beta'\sigma$) together with two tightly bound zinc atoms. The function of the σ subunit is to locate a promoter site where transcription is initiated. The σ subunit then dissociates from the rest of the enzyme, leaving the core enzyme ($\alpha_2\beta\beta'$) bound to the DNA template via the β' subunit. The β subunit carries the catalytic site for the internucleotide bond formation and is the target for antibiotic inhibition. This was demonstrated by studies with RNA polymerase isolated from rifampicin-resistant bacteria. Recent X-ray analysis of the holoenzyme from *Thermophilus aquaticus* revealed a DNA-binding channel in the active site cleft of the β subunit and an RNA exit channel.

Structural analysis of the RNA polymerase from *Escherichia coli* indicates that there are two distinct substrate binding sites on the β subunit. The 'i' or initiation site, is template-independent and recognizes only purine nucleoside triphosphates. The second site, called the i + 1 site, has no nucleotide preference. The initiation of transcription is marked by the formation of an internucleotide bond between the nucleotides bound to the i and i + 1 sites. Rifampicin and structurally related antibiotics have long been known to block initiation. The binding of the antibiotic is a two-stage process:

$$R + E \leftrightarrow RE \rightarrow RE^*.$$

The first stage is a fast bimolecular reaction, followed by a second, slower unimolecular process involving a conformation change in the enzyme which is necessary for the inhibitory action of rifampicin. The overall dissociation constant for the interaction is very low, 3 nm, indicating tight but not covalent binding. The crystal structure of the core of RNA polymerase (from *Thermophilus aquaticus*) complexed with rifampicin shows that the antibiotic binds in a pocket of the β subunit deep within the channel which accommodates the DNA–RNA hybrid but is more than 12Å from the active site. The napthol ring of the antibiotic forms van der Waals interactions with hydrophobic side chains of adjacent amino acids in the pocket. The analysis also shows the potential for hydrogen bonds between the five polar groups of rifampicin and neighbouring amino acids. Rifampicin inhibits the polymerase by sterically obstructing the path of the growing RNA chain after two, or at the most three, nucleotides have been added. The formation of the first internucleotide bond is not inhibited by rifampicin, and if chain growth progresses beyond the second or third phosphate diester bond before the addition of the drug, further chain elongation is insensitive to the action of rifampicin.

Another complex antibiotic, streptolydigin, is also a specific inhibitor of bacterial RNA polymerase. However, it inhibits chain elongation as well as the initiation process and increases the stability of purified RNA polymerase-DNA template complexes. The β subunit of the polymerase core enzyme bears the streptolydigin binding site, and the increased stability of the enzyme-template-antibiotic complex delays the progress of the enzyme along the template without affecting the accuracy of the transcriptional process.

Despite the evidence obtained *in vitro* for the mode of action of streptolydigin, studies of its effects on intact *Escherichia coli* cells indicate that *in vivo* streptolydigin may accelerate the termination of RNA chains. The rate of elongation of RNA chains is unaffected, but streptolydigin may destabilize the transcription complex *in vivo,* thus permitting premature attachment of termination factors. Streptolydigin has not found a clinical application.

4.6 Inhibition of nucleic acid synthesis by interaction with DNA

As we have seen, the synthesis of nucleic acids can be interrupted by blocking the supply of essential nucleotides, by the termination of chain extension following the incorporation of some nucleotide analogues into the nucleic acid structure, and by the direct inhibition of certain polymerases. The formation of a complex between the inhibitor and DNA, either covalent or noncovalent, may disrupt template function in replication and transcription. Some of the best-known examples of this group have useful anticancer activity and include actinomycin D, bleomycin and mitomycin C. These compounds are powerfully cytotoxic to mammalian and microbial cells alike, and therefore have no application in antimicrobial therapy and will not be further discussed. A possible role for intercalation in the inhibition of bacterial topoisomerases by quinolones has already been mentioned. Several other antimicrobial agents are known to intercalate into the double helix of DNA, although this may not always be the primary basis for their antimicrobial activity. The intercalation phenomenon also has some useful practical applications in the laboratory.

4.6.1 Acridines, phenanthridines and choroquine

The medical history of the acridine dyes extends over some 90 years since proflavine (Figure 4.13) was used as a topical disinfectant on wounds during the First World War. Proflavine is too toxic to be used as a systemic antibacterial agent, but the related acridine, mepacrine (Figure 1.2), found wide application as an antimalarial drug before its replacement by more 'patient friendly' drugs. Chloroquine (Figure 4.13) is still an important antimalarial agent although, as we shall see in Chapter 6, its interaction with DNA is no longer considered to be the primary basis of its action against the malarial parasite. The phenanthridine ethidium (Figure 4.13) has some application as a trypanocide in veterinary medicine. All these compounds are characterized by flat, planar fused ring systems which are the key to their intercalating property.

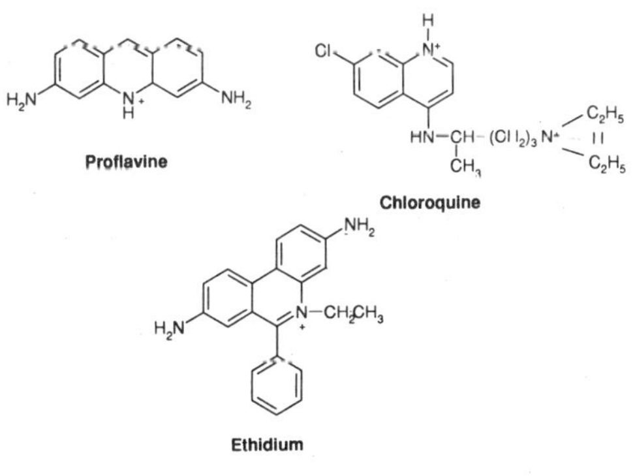

FIGURE 4.13 Three compounds that intercalate with DNA.

The compounds bind to the nucleic acids of living cells and the phenomenon forms the basis of the technique known as vital staining, since the nucleic acid-dye complexes exhibit characteristic colours when examined by fluorescence microscopy. The dyes also bind readily to nucleic acids *in vitro*, and the visible absorption spectra of the ligand molecules undergo a metachromatic shift to longer wavelengths.

Two types of binding to DNA are recognized: a strong primary binding which occurs in a random manner in the molecule, and a weak secondary binding. The strong primary binding occurs only with DNA, although many other polymers bind the dyes by the secondary process. The primary binding to DNA, which is mainly responsible for the ability of the drugs to inhibit nucleic acid synthesis, depends upon the intercalation of the rigidly planar molecules between the adjacent stacked base pairs of the double helix of DNA.

The evidence for this unusual type of interaction is provided by various measurable physical changes in the DNA:

1. Solutions of DNA show increased viscosity.
2. There is a decrease in the sedimentation coefficient of DNA determined by ultracentrifugation, which indicates a reduction in its buoyant density.
3. The thermal stability of DNA, i.e. the temperature at which the double helix begins to unwind or 'melt', is increased.

The extent of these changes is proportional to the amount of drug intercalated into the double helix. In the case of the anticancer drug actinomycin D, there is direct X-ray crystallographic evidence for intercalation into an oligo-deoxyribonucleotide. The DNA-compound complex is both straighter and stiffer than the uncomplexed nucleic acid and these changes raise the viscosity of solutions of DNA treated with intercalating agents. The reductions in sedimentation coefficient and buoyant density of DNA following intercalation result from a reduction in the mass per unit length of the nucleic acid. For example, a proflavine molecule increases the length of the DNA by about the same amount as an extra base pair, but because proflavine has less than half the mass of the base pair, the mass per unit length of the complexed DNA is decreased. The increased thermal stability of intercalated DNA is probably due in part to the extra energy needed to remove the bound compound from the double helix in addition to that required to separate the strands.

To permit the insertion of an intercalating molecule into DNA, it is believed that a local partial unwinding of the double helix associated with the normal molecular motions within the macromolecule produces spaces between the stacked base pairs into which the planar polycyclic molecule can move. A model advanced many years ago and still generally accepted shows schematically how polycyclic structures may intercalate between the stacked base pairs (Figure 4.14). The hydrogen bonding between the base pairs remains undisturbed, although there is some distortion of the smooth coil of the sugar phosphate backbone because the intercalated molecules maintain the double helix in a partially unwound configuration. It is believed that the distortion of the double helix, together with the hindrance to strand separation, are major factors in blocking DNA replication and transcription.

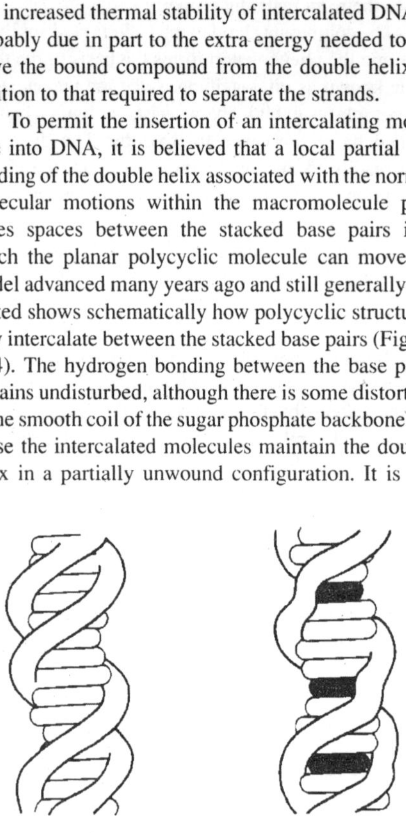

FIGURE 4.14 Diagram to represent the secondary structure of normal DNA (left) and DNA containing intercalated molecules (right). The stacked bases of the nucleic acid are separated at intervals by the intercalators (black), resulting in some distortion of the sugar-phosphate backbone of the DNA. [Reproduced with permission from L. Lerman (1964) *J. Cell Comp. Physiol.* **64**, Suppl. 1 (1964). Copyright owned by Wiley-Liss, Inc., a subsidiary of John Wiley & Sons.]

The details of specific drug–DNA interactions depend largely on the structures of the individual drugs. Intercalation does not involve the formation of covalent bonds between the compound and DNA. In general terms, the complex is probably stabilized by electronic interactions between the planar ring systems of the compounds and the heterocyclic bases of the DNA above and below the drug. The complexes formed by proflavine and ethidium may also be stabilized by hydrogen bonding between their amino groups and the charged oxygen atoms of the phosphate groups in the sugar-phosphate backbone. In the case of chloroquine, the projecting cationic side chain may form a salt link with a phosphate residue.

In certain tumour viruses and bacteriophages, in the kinetoplasts of trypanosomes and in bacterial plasmids (Chapter 8), double-stranded DNA exists as covalently closed circles. Circular DNA, covalently closed via the 3′-5′-phosphodiester bond, is characteristically supercoiled because the circular molecule is in a state of strain. The strain is relieved and the supercoils often disappear when single-stranded breaks or 'nicks' are produced by endonuclease action. Closed circular DNA has an unusual affinity for intercalating molecules which, because they partially unwind the double helix, also reduce the supercoiling of the DNA. If the unwinding proceeds beyond a certain point, as more and more drug is added the DNA begins to adopt the supercoiled form again, but in a direction opposite from that of the uncomplexed DNA. At this point the affinity of the closed circular DNA for the intercalated molecules declines until it is less than that of nicked DNA. The diminished affinity of closed circular DNA for ethidium at high concentrations of the drug permits a convenient separation of closed circular DNA from nicked DNA because the sedimentation coefficient and buoyant density of DNA with a lower content of intercalated compound are significantly higher. This effect is useful in the isolation of closed circular DNA on a preparative scale.

It is also possible that the initially higher affinity of supercoiled DNA for intercalating molecules may in part account for their specificity of action against organelles and organisms that contain circular DNA.

Thus the treatment of bacteria with acridines can induce the loss of plasmids from cells. The mitochondria of certain strains of yeast are severely and irreversibly damaged by growth in the presence of ethidium, apparently owing to drug-induced mutations which affect the mitochondrial DNA. The kinetoplast of trypanosomes is also seriously affected by intercalating agents, DNA synthesis in this organelle being selectively inhibited. Eventually the kinetoplast disappears altogether. Since this adversely affects the life cycle of trypanosomes, it is possible that the selective attack on the kinetoplast may underlie the trypanocidal activity of intercalating drugs such as ethidium.

Further reading

Angehrn, P. *et al.* (2004). New antibacterial agents derived from the DNA gyrase inhibitor cyclothialidine. *J. Med.Chem.* **47**, 1487.

Baca, A. M. (2000). Crystal structure of *Mycobacterium tuberculosis* 7,8, dihydropteroate synthase in complex with pterin monophosphate: new insight into the enzymatic mechanism and sulfa-drug action. *J. Mol. Biol.* **302**, 1193.

Campbell, E. A. *et al.* (2001). Structural mechanism for rifampicin inhibition of bacterial RNA polymerase. *Cell* **104**, 901.

Champoux, J. J. (2001). DNA topoisomerases: structure, function and mechanism. *Annu. Rev. Biochem.* **70**, 369.

Heddle, J. and Maxwell, A. (2002). Quinolone-binding pocket: role of GyrB. *Antimicrob. Agents Chemother.* **46**, 1805.

Hitchings, G. H. (1983). Inhibition of folate metabolism in chemotherapy. *Handb. Exp. Pharmacol.* **64**, 11.

Lewis, R. J. *et al.* (1996). The nature of inhibition of DNA gyrase by the coumarins and the cyclothialidines revealed by X-ray crystallography. *EMBO J.* **15**, 1412.

Maxwell, A. and Lawson, D. M. (2003). The ATP-binding site of Type II topoisomerase as a target for antibacterial drugs. *Curr. Topics Med. Chem.* **3**, 283.

Menendez-Arias, L. (2002). Targeting HIV: antiretroviral therapy and development of drug resistance. *Trends Pharmacol. Sci.* **23**, 381.

Rastelli G. *et al.* (2000). Interaction of pyrimethamine, cycloguanil, WR99210 and their analogues with *Plasmodium falciparum* dihydrofolate reductase: structural basis of antifolate resistance. *Bioorg. Med. Chem.* **8**, 1117.

Ren, J. *et al.* (1995) High-resolution structures of HIV-1 Reverse Transcriptase from four RT-inhibitor complexes. *Nat. Struct. Biol.* **2**, 293.

Rongbao, Li *et al.* (2000). Three-dimensional structure of *M. tuberculosis* dihydrofolate reductase reveals opportunities for the design of novel tuberculosis drugs. *J. Mol. Biol.* **295**, 307.

Tam, R. C. *et al.* (2002). Mechanisms of action of ribavirin in antiviral therapies. *Antivir. Chem. Chemother.* **12**, 261.

Vassylyev, D. G. *et al.* (2002). Crystal structure of a bacterial RNA polymerase holoenzyme at 2.6Å resolution. *Nature* **417**, 712.

Wilson, W. D. and Jones, R. L. (1981). Intercalating drugs: DNA binding and molecular pharmacology. *Adv. Pharmacol. Chemother.* **18**, 177.

Yuvaniyama, J. *et al.* (2003). Insights into antifolate resistance from malarial DHFR-TS structures. *Nat. Stuct. Biol.* **10**, 357.

Chapter five

Inhibitors of protein biosynthesis

The process of protein biosynthesis, in which the information encoded by the four-letter alphabet of nucleic acid bases is translated into defined sequences of amino acids linked by peptide bonds, is an exquisitely complex process involving more than 100 macromolecules. Amino acid-specific transfer RNA molecules, messenger RNAs and many soluble proteins are required, in addition to the numerous proteins and three types of RNA that make up the ribosomes. Although many general features of the protein synthetic machinery are similar in prokaryotic and eukaryotic organisms, several naturally occurring compounds and currently one series of wholly synthetic compounds specifically inhibit bacterial protein synthesis and provide us with drugs of considerable therapeutic value. It is intriguing, therefore, that 'nature' has been much more successful than synthetic organic chemistry in producing compounds that discriminate between bacterial and mammalian protein synthesis. Inhibitors specific for protein synthesis in fungi are as yet unknown, probably because the similarities between the mechanisms in fungal and mammalian cells are too close to permit this degree of discrimination. We provide a brief outline of ribosomal structure and the sequence of events in protein biosynthesis. The subsequent discussion is mainly concerned with the modes of action of inhibitors of bacterial protein synthesis in clinical use.

5.1 Ribosomes

These remarkable organelles are the machines upon which polypeptides are elaborated. There are three main classes of ribosomes, identified by their sedimentation coefficients in the ultracentrifuge. The 80S ribosomes are confined to eukaryotic cells, while 70S ribosomes are characteristic of prokaryotic cells. A species of 50–55S ribosome in mammalian mitochondria resembles bacterial ribosomes in functional organization and antibiotic sensitivity; analogous small ribosomes also occur in the chloroplasts of green plants. The 80S particle dissociates reversibly into 60S and 40S subunits, and the 70S ribosome into 50S and 30S subunits, when the Mg^{2+} concentration of a suspending solution is reduced. Both 80S and 70S ribosomes are composed exclusively of protein and RNA in mass ratios of approximately 50:50 and 35:65, respectively. There are three distinct species of RNA in most ribosomes, with sedimentation coefficients of 29S, 18S and 5S in 80S particles from animal cells; 25S, 18S and 5S in 80S particles from plant cells; and 23S, 16S and 5S in 70S particles. The 55S ribosomes contain two RNA species that sediment at about 16S and 12S, but probably no 5S RNA. The protein composition of ribosomes is impressively complex. The 30S subunit of *Escherichia coli* ribosomes contains 21 proteins ('S' proteins) and the 50S ('large') subunit contains 34 proteins ('L' proteins). The amino acid sequences of all of these proteins are now known.

Great progress has been made in understanding how the ribosome is constructed, and a consensus model of the gross structures of the subunits of the 70S ribosome, based upon electron microscopic images, is represented in simplified form in Figure 5.1. Furthermore, success in obtaining high-quality crystals of ribosomes from certain bacteria, including *Thermus thermophilus, Deinococcus radiodurans* and *Haloarcula marismortui* (an archaeobacterium) has facilitated high-resolution X-ray analyses of the two ribosomal subunits (Figure 5.2) and their interactions with several antibiotics of major medical importance. Although none of these bacteria are pathogenic, the generality of the 70S ribosomal structure makes it reasonable to conclude that the data from their ribosomes are applicable to those of pathogenic bacteria, especially those from *Deinococcus radiodurans,* which are more sensitive to antibiotics than the archaeobacterium ribosomes. It has proved more difficult to crystallize ribosomes from bacteria such as *Escherichia coli* or *Staphylococcus aureus* because these ribosomes tend to deteriorate rapidly under the conditions optimal for crystal formation. However, a lower resolution (between 9 and 10Å) structure of the 70S *Escherichia coli* ribosome is now available. The detailed and complex crystallographic data are beyond the scope of this book, and the interactions between ribosomes and their antibiotic inhibitors will therefore be discussed in outline only. Detailed descriptions are available in references listed under 'Further reading.'

There are two main functional regions in ribosomes, known as the translational and exit domains. Both ribosomal subunits contribute to the translational domain; mRNA binding and its interaction with aminoacyl-tRNAs, i.e. decoding, occur on the 30S subunit in the region called the platform, whereas peptide bond formation is catalyzed by the peptidyl transferase activity located on the central protuberance of the 50S subunit. The lengthening peptide chain leaves the 50S subunit via the exit domain which is found on the side of the subunit opposite to the peptidyl transferase site. The topographies of the various elongation and initiation factor binding sites are also well understood and represent the culmination of many years of effort by a number of groups using highly sophisticated techniques. Similar features have been recognized in eukaryotic ribosomes, which bind to the rough endoplasmic reticulum at the exit site.

For many years it was believed that ribosomal proteins played the leading role in decoding mRNA and in peptide bond synthesis. However, it is now clear that ribosomal RNA is critically involved in decoding (16S rRNA) and in peptide bond synthesis (23S rRNA), with ribosomal proteins being required to maintain the functional three-dimensional structures of the RNA molecules. As we shall see, these discoveries have major implications for the sites of action of several antibiotics that inhibit protein biosynthesis.

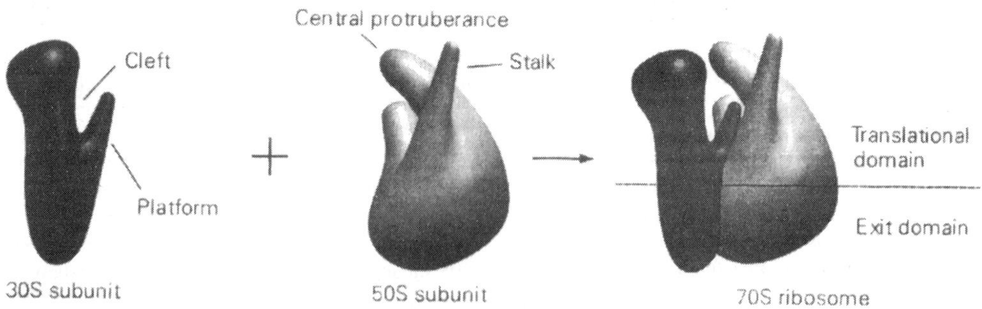

FIGURE 5.1 Simplified representation of the subunits of a prokaryotic ribosome and their cooperative interaction to form the functional 70S particle based upon images obtained by electron microscopy. (This diagram was kindly provided by Paul J. Franklin.)

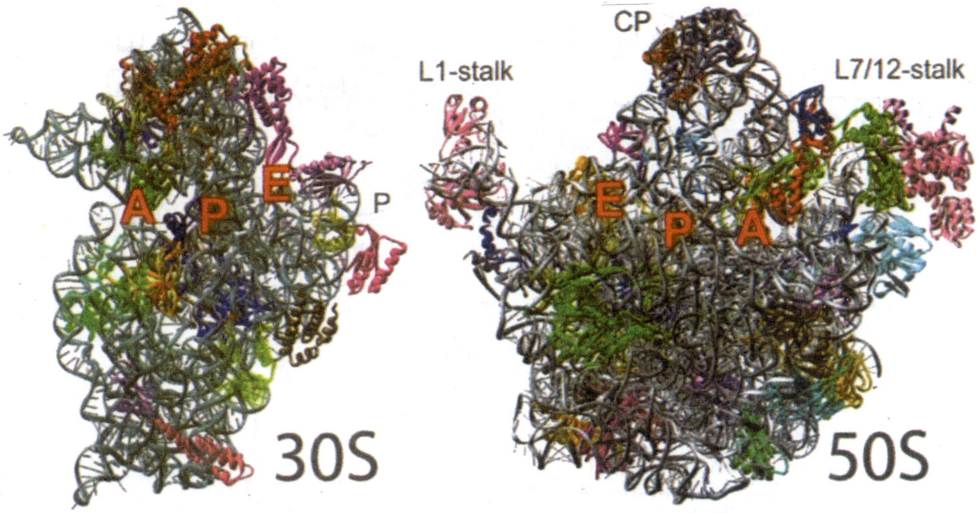

FIGURE 5.2 High-resolution structures resolved by X-ray crystallographic analysis of the two ribosomal subunits. The 30S subunit was crystallized from *Thermus thermophilus* and the 50S subunit from *Deinococcus radiourans*. The RNA chains are represented as blue-gray (30S) or gray (50S) ribbons and the protein main chains are shown in different colors. CP, central protuberance; P, platform. A, P and E (in red) indicate the sites where tRNA molecules bind to the large and small subunits. [Reproduced with permission from T. Auerbach *et al. Current Drug Targets. Infectious Disorders* 2, 169–176 (2002).]

5.2 Stages in protein biosynthesis

5.2.1 Formation of aminoacyl-transfer RNA

Each amino acid is converted by a specific aminoacyl-tRNA synthetase to an aminoacyladenylate which is stabilized by association with the enzyme:

$$\text{ATP} + \text{amino acid (aa)} \xleftrightarrow{\text{Aminoacyl-tRNA synthetase}} \text{aa-AMP-Enz} + \text{PP}_i.$$

Each amino acid-adenylate-enzyme complex then interacts with an amino acid-specific tRNA to form an aminoacyl-tRNA in which the aminoacyl group is linked to the 3′-OH ribosyl moiety of the 3′ terminal adenosyl group of the tRNA by a highly reactive ester bond:

$$\text{aa-AMP-Enz} + \text{tRNA} \leftrightarrow \text{aminoacyl-tRNA} + \text{AMP} + \text{ENZ}$$

The subsequent stages in prokaryotic protein biosynthesis are outlined in Figure 5.3.

5.2.2 Initiation

Three protein factors, IF1, IF2 and IF3, loosely associated with 70S ribosomes are concerned with initiation. IF1 enhances the rate of ternary complex formation between mRNA, initiator tRNA and 30S ribosomal subunits. IF1 also has a role in promoting the dissociation of 70S ribosomes released from previous rounds of polypeptide synthesis into 30S and 50S subunits. Factor IF3, which then binds to the 30S subunit, is also needed for the binding of mRNA. The complex containing the 30S subunit, IF3 and mRNA is joined by IF2, GTP and the specific initiator tRNA, *N*-formylmethionyl-tRNA_F (fMet-tRNA_F), the role of IF2 being to direct the binding of fMet-tRNA_F to a specific initiator codon, usually AUG but occasionally GUG. IF1 and IF2 are now ejected from the complex, a process dependent on the hydrolysis of one molecule of GTP to GDP and inorganic phosphate. The next stage involves the detachment of IF3 in the presence of a 50S subunit to permit the formation of the 70S ribosome.

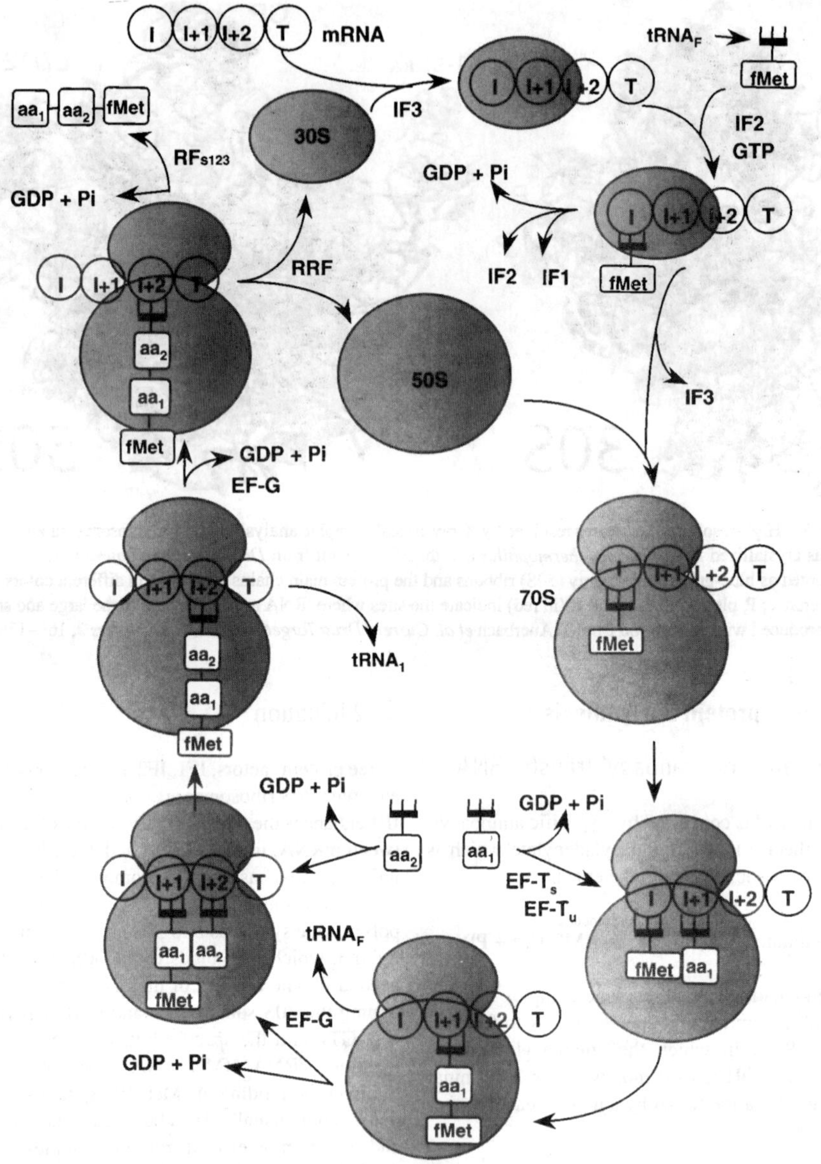

FIGURE 5.3 Diagrammatic summary of the main stages in the biosynthesis of proteins on 70S ribosomes. The scheme should be read clockwise starting at the top. I, I + 1, I + 2 represent the initiator and successive codons; T is the terminator codon on mRNA; fMet, aa_1, aa_2, are *N*-formylmethione and two other amino acids; $tRNA_F$, $tRNA_1$, $tRNA_2$, are specific transfer RNAs. The involvement of the various protein cofactors referred to in the text is also indicated.

The association of the 50S and 30S subunits is believed to involve interactions between their protein and RNA chains. Initiation on 80S ribsomes resembles that on 70S ribsomes except that eukaryotic initiation *in vivo* uses unformylated Met-tRNA^{Met}. In addition, there are at least nine eukaryotic initiation factors whose interplay is much more complex than that in prokaryotic organisms.

5.2.3 Peptide bond synthesis and chain elongation

The consensus view of synthetic sequence rests largely on the concept of three distinct sites on the ribosome, called the acceptor (A) site, the donor or peptidyl (P) site and the exit (E) site. The A site is the primary decoding site where the codon of the mRNA first interacts with the anticodon region of the specific aminoacyl-tRNA. The fMet-tRNA$_F$, however, binds directly to the P site. The binding of the next aminoacyl-tRNA to the A site requires protein factors EF-T$_s$ and EF-T$_u$. EF-T$_u$ binds GTP and then forms a ternary complex with aminoacyl-tRNA. This complex binds to the acceptor site, with accompanying hydrolysis of one molecule of GTP. GTP hydrolysis is not essential for the binding of aminoacyl-tRNA, but in its absence the bound aminoacyl-tRNA is not available for peptide bond formation. The role of the stable factor, EF-T$_s$, is to regenerate EF-T$_u$-GTP from EF-T$_u$-GDP by stimulating the exchange of bound GDP for a molecule of free GTP. Apparently EF-T$_s$ forms a high-affinity intermediate complex with EF-T$_u$, and GDP is lost from this intermediate. An important and characteristic feature of the ribosome is the fidelity with which it translates the genetic code of the mRNA into the correct amino acid sequence. The interaction energy of base pairing leads readily to mismatching of codon and anticodon. Ribosomes, however, reject mismatches with an error frequency as low as 10^{-4}. The binding of the ternary complex of aminoacyl-tRNA-EF-Tu-GTP to a cognate mRNA codon at the A site results in a conformational change in the ribosome from an 'open' state to the 'closed' state, which then drives the irreversible chemical changes leading to peptide bond formation. In the event of a mismatch between the codon of mRNA and the tRNA anticodon, the conformational change in the ribosome does not occur and it remains in the open state. The mismatched aminoacyl tRNA then dissociates from the A site. As we shall see, these events are critical to an understanding of the miscoding activity of aminoglycoside antibiotics.

After correct matching of aminoacyl tRNA and the mRNA codon at the A site, the scene is set for the formation of the first peptide bond. The carboxyl group of the *N*-formylmethionine attached to the P site through its tRNA is 'donated' to the amino group of the adjacent amino acid at the A site to form a peptide bond. The formation of the peptide bond is catalyzed by peptidyl transferase, which is a complex component of the 50S subunit located at the central protuberance of the 50S subunit. Although several ribosomal proteins, including L2, L15, L16 and L27, are located in the vicinity of the peptidyl transferase site, the current view is that 23S rRNA is responsible for catalyzing peptide bond formation. The reader is referred to the relevant papers listed under 'Further reading' for details of the complex structures of both 16S and 23S rRNA. Suffice it to say that the peptidyl transferase function is probably located in the region of the 23S rRNA known as domain V, which also harbors the binding site for protein L27. The dipeptide which is formed at domain V remains attached through its C-terminal to the second tRNA at the A site. The dipeptidyl tRNA is then translocated from the A to the P site, still linked to the mRNA (through the codon–anticodon interaction). The third consecutive codon of the mRNA is now exposed at the A site by the relative movement of the ribosome towards the 3′ end of the mRNA. The translocation step requires factor EF-G and the hydrolysis of another molecule of GTP. EF-G binds to the L7/L12 region of the large subunit. This area is implicated in GTP hydrolysis mediated by EF-T$_u$, EF-G and IF2. After peptide bond formation, the deacylated tRNA transfers to the E site, which promotes its ejection from the ribosome. It is not clear whether there is base pairing at the E site between the mRNA codon and the anticodon of the tRNA. The growing polypeptide chain passes into and through an exit tunnel which stretches approximately 100Å from the peptidyl transferase site through the ribosome to the point of exit.

5.2.4 Chain termination and release

The signal for termination of the polypeptide chain is given by the appearance of one of three terminator codons—UAA, UAG or UGA—at the A site. The complete polypeptide is detached from the tRNA at the C-terminal amino acid, a step that requires peptidyl transferase activity and the release factors RF1, RF2 and RF3. Factors RF1 and RF2 are concerned with the recognition of specific terminator codons, RF1 recognizing UAA and UAG and RF2, UAA and UGA. Both the binding of RF1 and RF2 to the ribosomes and their release require RF3, which also promotes the cleavage of peptidyl-tRNA to release the completed protein chain. Hydrolysis of GTP is also involved in the release reaction. Release from eukaryotic ribosomes involves only one codon-recognizing release factor and requires the cleavage of GTP. The formyl groups of the fMet ends of prokaryotic polypeptides are removed by a specific enzyme and in many proteins the methionine residue is also removed. After release of the completed polypeptide, the ribosome is liberated from the mRNA and deacylated tRNA by the combined action of GTP, EF-G and ribosome release factor (RRF), which permits dissociation into 30S and 50S subunits. IF3 then binds to the 30S subunit. This prevents reassociation until a full initiation complex has once more been completed.

5.3 Puromycin

The antibiotic puromycin is a unique inhibitor of protein biosynthesis, since the drug itself reacts to form a peptide with the C-terminal of the growing peptide chain on the ribosome, thus prematurely terminating the chain. This remarkable property gave puromycin an important role in the elucidation of the mechanism of peptide bond formation and, as we shall see, in defining the point of action of several other inhibitors of protein biosynthesis.

The structural similarity of puromycin to the terminal aminoacyladenosine moiety of tRNA was noted many years ago (Figure 5.4) and this proved to be the key to understanding its actions. Aminoacyladenosine is the terminal residue of tRNA in both prokaryotic and eukaryotic organisms. Puromycin therefore terminates protein synthesis equally effectively on 70S and 80S ribosomes and the antibiotic has no therapeutic value. The structural analogy of puromycin with aminoacyladenosine led to the demonstration that the amino group of the antibiotic forms a peptide bond with the acyl group of the terminal aminoacyladenosine moiety of peptidyl-tRNA attached to the ribosome. No further peptide bond formation is possible because of the chemical stability of the C—N bond which links the *p*-methoxyphenylalanine moiety of puromycin to the nu-

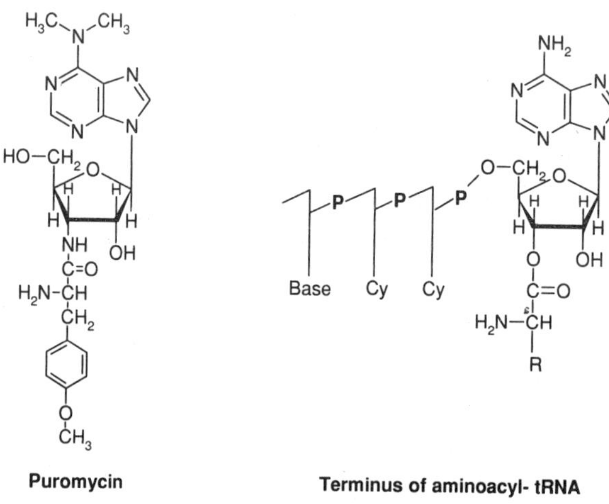

FIGURE 5.4 Structural analogy between puromycin and the aminoacyl terminal of transfer RNA. Cy represents cytosine and R represents the rest of the amino acid molecule.

cleoside residue. Peptidyl-puromycin then dissociates from the ribosome.

Provided that the peptidyl-tRNA is in the P site on the ribosome, its reaction with puromycin has no other requirement than a normally functioning peptidyl transferase activity. Puromycin does not, however, react with peptidyl-tRNA at the A site; in this situation factor EF-G and GTP must be added in order to effect translocation of the peptidyl-tRNA to the P site. Only then is peptidyl-puromycin formed and released from the ribosome. The puromycin reaction occurs fairly readily at 0° C, whereas normal chain elongation is negligible at this temperature, suggesting that puromycin has a considerable competitive advantage over aminoacyl-tRNA in reacting with the peptidyl-tRNA. The puromycin reaction will in fact proceed under greatly simplified conditions, requiring only 50S ribosomal subunits, the oligonucleotide CAACCA-(fMet) to replace the peptidyl tRNA normally found at the P site, and Mg^{2+} and K^+ ions. This simple system, known as the fragment reaction, allows the separation of peptide bond formation from the more complex process of translation. It has been extremely useful in the investigation of those antibiotics suspected of inhibiting peptide bond synthesis.

Derivatives of puromycin indicate that a single benzene ring in the side chain is necessary for activity; replacement of the *p*-methoxyphenylalanine with proline, tryptophan, benzylhistidine or any aliphatic amino acid results in a marked loss of activity. The L-phenylalanine analogue is about half as active as puromycin, while the D-phenylalanine analogue is completely inactive. Replacing the *p*-methoxyphenylalanine residue with the *S*-benzyl-L-cysteine analogue results in only a minor loss of activity, which may be due to the increased distance between the benzene ring and the free NH_2 group caused by the additional S and C atoms. Since puromycin substitutes for all aminoacyl-tRNAs equally well, the sufficiency of a single benzene ring in the amino acid moiety of puromycin and its analogues is puzzling. The aromatic ring may be involved in a hydrophobic interaction with the terminal adenosine of peptidyl-tRNA at the donor site, thus contributing to the formation of an intimate complex between puromycin and peptidyl-tRNA prior to the formation of a peptide bond. In view of the structure of the aminoacyladenosine of the tRNA terminal, the requirement for linkage of the amino acid moiety to the ribose 3′ position of puromycin is, however, not unexpected. Puromycin substituted in the 5′ position of the ribose with cytidylic acid is an effective peptide chain terminator, and there is an absolute requirement for cytidine in this derivative. Presumably this substitution extends the structural analogy with tRNA.

5.4 Inhibitors of aminoacyl-tRNA formation

Several naturally occurring and synthetic analogues of amino acids inhibit the formation of the aminoacyl-tRNA complex. Close analogues may become attached to the appropriate tRNA and subsequently become incorporated into abnormal proteins. Among these are ethionine, norleucine, *N*-ethylglycine and 3,4-dehydroproline. Several naturally occurring antibiotics, such as borrelidin, furanomycin and indolmycin, competitively antagonize the incorporation of the corresponding amino acids, i.e. threonine, isoleucine and tryptophan, respectively, into aminoacyl-tRNA. Most of these inhibitors of aminoacyl-tRNA formation lack specificity against micro-organisms and hence have no useful medical application, although indolmycin is said to be specific for prokaryotic tryptophanyl-tRNA synthetase. However, by far the most important inhibitor of aminoacyl-tRNA synthesis is the antibiotic mupirocin or pseudomonic acid A (Figure 5.5), which is produced by *Pseudomonas fluorescens*. Mupirocin has excellent activity against several species of *Staphylococcus*, and is especially useful against the dangerous methicillin-resistant *Staphylococcus aureus* (MRSA). The antibiotic has limited activity against Gram-negative bacteria but includes in its spectrum

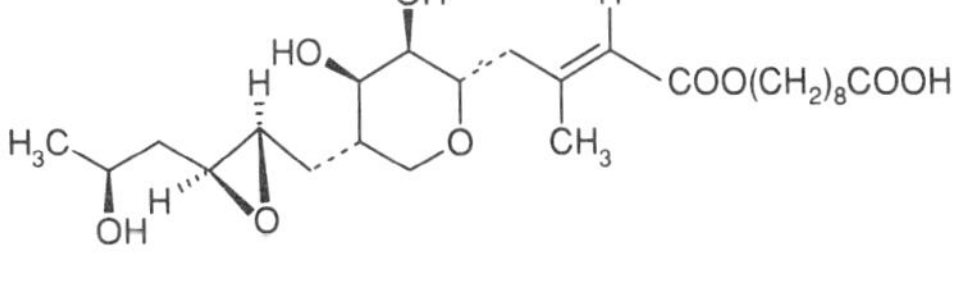

FIGURE 5.5 Mupirocin (pseudomonic acid A).This antibiotic, which has useful topical activity against methicillin-resistant *Staphylococcus aureus* and *Streptococcus pyogenes*, is a specific inhibitor of bacterial isoleucyl-tRNA synthetase.

Haemophilus influenzae, Neisseria gonorrhoeae and *Neisseria meningitidis.* Unfortunately, although mupirocin is well absorbed after oral doses, it is rapidly metabolized in the body to the inactive monic acid and its clinical use is therefore confined to topical applications. It is effective in the treatment of the skin infection impetigo caused by *Staphylococcis aureus* and *Streptococcus pyogenes.* Mupirocin is also useful in eliminating MRSA from the nasal passages of hospital staff and vulnerable patients.

The antibacterial activity of mupirocin depends upon its specific inhibition of isoleucyl-tRNA synthetase. Biochemical data indicate that the compound is a competitive inhibitor of the isoleucyl adenylate although the chemical structures are significantly different. However, X-ray crystallography of isoleucyl-tRNA synthetase from *Thermus thermophilus* shows that the antibiotic binds to the catalytic cleft of the enzyme, confirming that it does indeed compete with isoleucyl adenylate. The antibacterial specificity of mupirocin, which rests on the fact that the compound has little or no inhibitory activity against eukaryotic isoleucyl-tRNA synthetase, is apparently explained by differences in just two amino acids of the bacterial and eukaryotic enzymes. First, histidine-581 in the bacterial enzymes is replaced by an asparagine or serine in eukaryotes. Second, leucine-583 in *Thermus thermophilus*, or phenylalanine-583 in many other bacterial species, is replaced by isoleucine in the eukaryotic enzymes. The crystallographic data suggest that the presence of isoleucine in eukaryotic isoleucyl-tRNA synthetase markedly weakens the binding of mupirocin and renders the antibiotic ineffective as an inhibitor of the human form of the enzyme.

5.5 Inhibitors of initiation and translation

5.5.1 Streptomycin

This naturally occurring antibiotic is a member of the aminoglycoside group and has the complex chemical structure illustrated in Figure 5.6. Streptomycin was discovered in the early 1940s and was the first drug really effective against tuberculosis, although it is less commonly used in the treatment of this disease today. It is a broad-spectrum antibiotic, active against a range

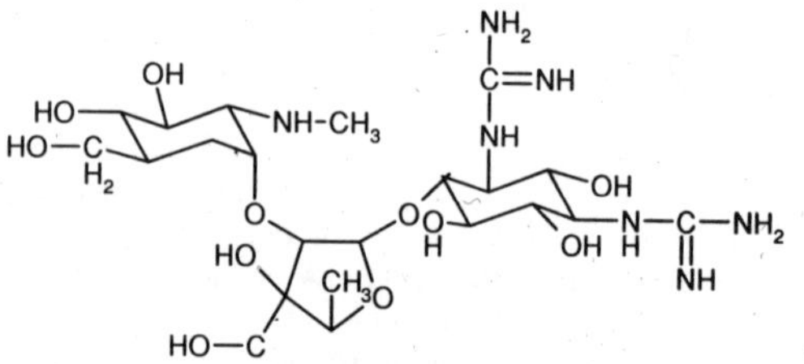

FIGURE 5.6 The aminoglycoside streptomycin, the first effective antitubercular drug.

of Gram-positive and Gram-negative bacteria, but its use is limited by several problems. First, the drug is effective only when given by injection because its absorption from the gastrointestinal tract is very poor. Second, along with other aminoglycosides, streptomycin may cause permanent deafness, owing to irreversible injury to the eighth cranial nerve, and may also cause kidney damage, although fortunately the latter is usually reversible. Third, bacterial resistance to this antibiotic develops readily.

Unusually for an inhibitor of protein biosynthesis, streptomycin is bactericidal rather than bacteriostatic. Cell death is preceded by marked effects on protein biosynthesis which are specific for the 70S ribosomes of bacteria:

1. Streptomycin strongly inhibits the initiation of peptide chains. The drug also slows the elongation of partly completed chains, although even at high concentrations of streptomycin, chain elongation is not completely suppressed. Peptidyl transferase activity is unaffected. These effects on initiation and elongation are attributed to a disturbance of the functions of both A and P sites by streptomycin.
2. Studies carried out on the effects of streptomycin in cell-free systems from bacteria using synthetic polynucleotides as messengers clearly show that streptomycin induces codon misreading. Thus the antibiotic inhibits the incorporation into peptide linkages of phenylalanine directed by poly(U) histidine and threonine directed by poly(AC), and arginine and glutamic acid directed by poly(AG).

In contrast, streptomycin may, under some conditions, stimulate the incorporation of amino acids in the presence of synthetic messengers which do not normally code for these amino acids. For example, while streptomycin inhibits the incorporation of phenylalanine in the presence of poly(U), the incorporation of isoleucine and serine is stimulated. Streptomycin also induces poly(C) to promote the incorporation of threonine and serine instead of proline. All these observations indicate that streptomycin distorts the proofreading selection of the correct aminoacyl-tRNA by the ribosome. However, the misreading is not random and the following rules are more or less observed:

1. In any mRNA codon, only one base is misread, usually a pyrimidine located at the 5′ end or middle position of the codon.
2. There is no misreading of the base at the 3′ end.
3. Misreading of purines is rare and the occurrence of these in a codon decreases the chance of misreading the codon.

The induction of misreading of the genetic message by streptomycin probably underlies the well-known ability of this antibiotic to suppress certain bacterial mutations.

Site of action of streptomycin

Streptomycin binds tightly but not irreversibly to 70S ribosomes, with a K_d of 10^{-7} M. There is also low-affinity binding ($K_d > 10^{-4}$ M), which is probably irrelevant to the action of streptomycin. For many years it was thought that the ribosomal binding site for streptomycin had been identified by experiments with bacterial mutants highly resistant to the antibiotic. Ribosomes prepared from streptomycin-sensitive and streptomycin-resistant *Escherichia coli* were dissociated into 30S and 50S subunits by lowering the Mg^{2+} concentration in the medium. A ribosomal subunit 'cross-over' experiment showed that reassociated 70S particles composed of 30S subunits from resistant cells and 50S subunits from sensitive cells were resistant to streptomycin. In the opposite cross, i.e. 30S subunits from sensitive cells and 50S subunits from resistant cells, the resulting 70S ribosomes were streptomycin sensitive. This indicated that the target site of streptomycin was on the 30S subunit, a view strengthened by the finding that radiolabelled streptomycin bound specifically to the 30S subunit but not to the 50S subunit of sensitive ribosomes. Streptomycin did not bind to the 30S subunit from resistant cells and did not induce misreadings of mRNA translated with resistant ribosomes. It was then found that in the resistant ribosomes, protein S12 was mutated. Several resistant variants were isolated with various single amino acid replacements. Thus lysine-42 was replaced by asparagine, threonine or arginine, while in another mutant lysine-87 was replaced by arginine alone. However, studies with 30S subunits treated with protein extractants to remove the S12 protein showed that it is not essential for protein synthesis nor is it an absolute requirement for streptomycin binding.

The accumulating evidence for a central role for ribosomal RNA (rRNA) in protein biosynthesis has been accompanied by the discovery of the importance of interactions between rRNA and several antibiotics and their ability to interfere with the biosynthetic process. Until recently, the crucial technique used to reveal the significance of rRNA was 'footprinting'. This detects the ability of antibiotics bound to ribosomes to protect the bases of specific nucleotides of 16S and 23S rRNA against chemical modification, usually alkylation, by reagents such as dimethyl sulfate. Such protection is considered to be evidence of specific interactions or binding between the antibiotics and functionally significant domains in rRNA which would otherwise be attacked by the chemically reactive reagent. When streptomycin binds to 70S ribosomes in the presence of dimethyl sulfate, electrophoretic analysis of 16S rRNA following its hydrolysis to individual nucleotides shows that the antibiotic affords protection for the bases of nucleotides 911 to 915, although protection is incomplete even when the streptomycin–ribosome binding is fully saturated. Useful as the footprinting method has been in the study of ribosome–antibiotic interactions, it has now largely been superseded by the remarkable insights achieved by X-ray crystallography. In the case of the aminoglycosides, the crystal structure of the 30S subunit from *Thermus thermophilus* complexed with streptomycin, paromomycin and spectinomycin has been solved to 3Å . The analysis shows that streptomycin is tightly bound to the phosphate backbone of helix-44 of the 16S rRNA, a major component of the A site, through salt bridges and hydrogen bonds. The four nucleotides implicated in streptomycin binding by footprinting were confirmed by the X-ray data. There are additional interactions with helix-27 and with lysine-45 of the S12 protein. It is now considered that streptomycin binds tightly to the A site on the 30S subunit, where it stabilizes the closed conformation of that site. X-Ray crystallography shows that streptomycin increases the number of interactions between the ribosomal shoulder and the central part of the 30S subunit. By promoting stabilization of the closed state of the ribosome, which would normally only be achieved by correct matching of codon and anticodon, the antibiotic permits mismatching to occur, with resultant misreading of the genetic code. The wild-type S12 protein contributes to the stability of the three-dimensional structure of the A site on 16S rRNA. Mutations in S12 that confer resistance to streptomycin no longer permit the antibiotic to exert its miscoding action at the A site.

Bactericidal action of streptomycin

Streptomycin and structurally related aminoglycosides are unusual among inhibitors of protein biosynthesis in causing the death of bacteria. Most other inhibitors merely arrest bacterial growth, which resumes when the antibiotic is removed from the microbial environment. It seems likely that the ability of the aminoglycosides to induce ribosomal misreading of mRNA is an important factor in their bactericidal action. The resulting aberrant proteins probably cause a variety of disordered activities within the cell, including disruption of the normal functions of the cytoplasmic membrane and the outer membrane of Gram-negative bacteria, leading to irreversible changes in cellular metabolism.

5.5.2 Other aminoglycoside antibiotics

In addition to streptomycin, there are several other aminoglycosides that are useful in antibacterial chemotherapy. These include neomycin, kanamycin, gentamicin (Figure 5.7), tobramycin, amikacin, and netilmicin, the last two of which are semisynthetic modifications of naturally occurring antibiotics designed to minimize enzymic inactivation by resistant bacteria (Chapter 9). Like streptomycin, these antibiotics are usually given by injection and are valuable in the treatment of serious Gram-negative infections. Paromomycin is unusual in finding its primary use against amebic infections of the intestinal tract.

Several of these antibiotics have effects on protein biosynthesis that differ from those of streptomycin. For example, gentamicin, kanamycin and neomycin exhibit three separate concentration-dependent effects on isolated ribosomes:

1. At concentrations below 2 $\mu g\ ml^{-1}$ there is strong inhibition of total protein synthesis associated with inhibition of the initiation step, but little induction of mRNA misreading.
2. Between 5 and 50 $\mu g\ ml^{-1}$ there is misreading, especially by reading through the termination signals. Protein synthesis may therefore actually increase through the accumulation of abnormally long polypeptides as the ribosomes continue past the end of one message and on to the next.
3. Higher antibiotic concentrations re-establish inhibition of protein synthesis.

Each ribosomal subunit has one strong binding site for kanamycin as well as a number of weak binding sites. Gentamicin and neomycin compete with kanamycin, suggesting similar binding domains which must be distinct from the streptomycin-binding region since there is no comparable inhibition with streptomycin. This conclusion is supported by footprinting studies with neomycin, kanamycin and gentamicin which indicate a pattern of protection different from that of streptomycin. Neomycin, kanamycin and gentamicin protect adenine (A)-1408, guanine (G)-1491 and G-1494 of 16S rRNA from chemical modification by dimethyl sulfate. These bases are located close to the decoding region of the 3′ end of 16S rRNA, and two of them (A-1408 and G-1494) are also protected from chemical probing by tRNA bound to the ribosomal A site. The data are consistent, therefore, with the miscoding effects of neomycin, kanamycin and gentamicin. The basis of the different concentration-dependent effects of the antibiotics is uncertain but may be associated with progressive sat-

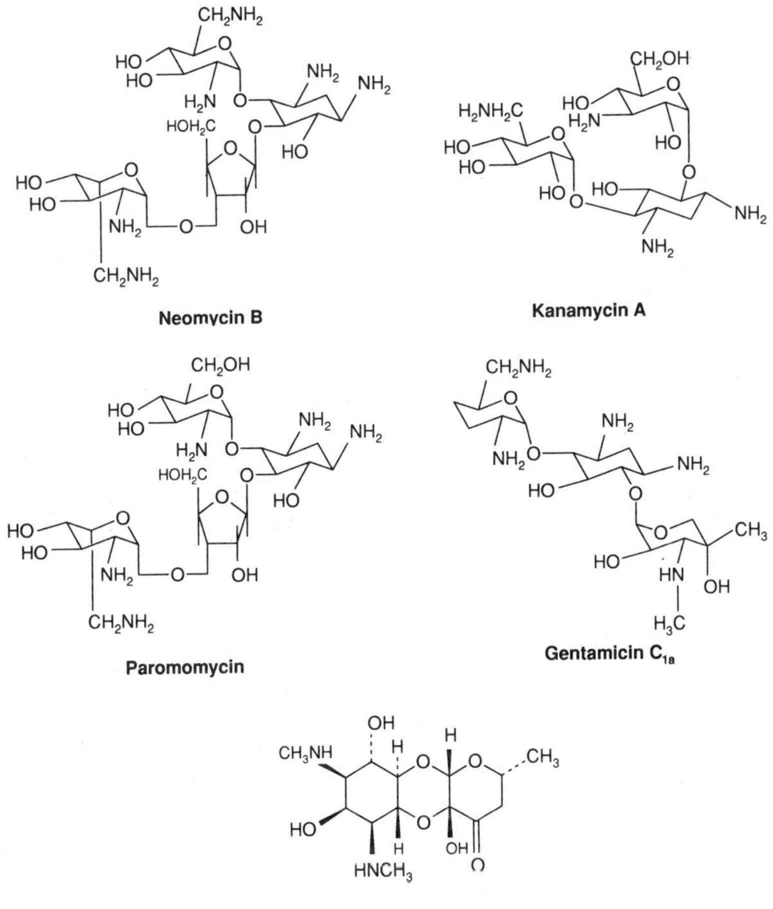

FIGURE 5.7 Examples of other aminoglycoside antibiotics. Spectinomycin is more appropriately described as an aminocyclitol antibiotic because it contains an inositol ring with two of its OH groups substituted by methylamino groups.

uration of their ribosomal binding sites with increasing concentration.

Crystallographic analysis of the complex of paromomycin and the 30S subunit from *Thermus thermophilius* reveals that the drug binds in the major groove of helix-44 of 16S rRNA. A nuclear magnetic resonance study of the interaction of gentamicin with a fragment of helix-44 indicated that it binds in much the same way as paromomycin. This probably applies to the other aminoglycosides as well. The binding to helix-44 accords with the miscoding activity of these aminoglycosides. X-Ray crystallography has revealed that subtle distinctions between their effects on miscoding and those of streptomycin are due to differences of detail in their various binding interactions with 16S rRNA.

Spectinomycin is usually included in the aminoglycoside group, even though it lacks an amino sugar residue (Figure 5.7). Unlike the previously mentioned aminoglycoside antibiotics, its action is bacteriostatic rather than bactericidal. The effects of spectinomycin on protein synthesis are also markedly different from those of the other aminoglycosides. While it inhibits protein synthesis in bacterial cells and in cell-free systems containing 70S ribosomes, spectinomycin does not induce misreading of mRNA. Spectinomycin has no effect on codon recognition, peptide bond formation, or chain termination and release. The antibiotic inhibits the translocation of peptidyl-tRNA from the A site to the P site. The rigidity of the spectinomycin molecule, conferred by the fused ring system, is thought to be relevant to the inhibitory action. The X-ray analysis of the spectinomycin-30S subunit complex referred to earlier shows that the drug binds in the minor groove of helix-34 of 16S rRNA, making a single contact with a 2′-hydroxyl group and hydrogen bond links with several other bases, especially with G-1064 and cytosine(C)-1192. Its binding mode is thus considerably different from the other aminoglycosides. The rigid spectinomycin molecule may sterically hinder a movement of the head region of the 30S subunit believed to be involved in the normal translocation event. With the emergence of β-lactamase-producing *Neisseria gonorrhoeae*, spectinomycin has become useful in the treatment of gonococcal infections.

Postscript: do aminoglycosides inhibit *trans*-translation in bacteria?

The recently discovered phenomenon of *trans*-translation rescues stalled ribosomes and contributes to the degradation of incompletely synthesized polypeptides. *Trans*-translation is unique to prokaryotes and is essential for the viability of some bacteria, for example, the important pathogen *Neisseria gonorroeae*. In this process a species of messenger RNA, termed transfer messenger RNA (tmRNA, or SsrA RNA or 10Sa RNA), acts first as tRNA being aminoacylated at the 3′ end with alanine and catalyzed by alanyl-tRNA synthetase. An alanine residue is thereby added to the end of the stalled protein chain. The tmRNA then reverts to its messenger function and translation resumes, not on the original messenger RNA, but at an internal position in the tmRNA. The arrested protein synthesis is released, ribosome recycling resumes and the previously stalled protein chain is now tagged at the C-terminal with a sequence which signals its subsequent degradation. Experiments *in vitro* show that several aminoglycosides, neomycin B being the most potent, block *trans*-translation by binding to the tRNA domain of tmRNA. This interaction disrupts the normal conformation of tmRNA and renders it incapable of efficient aminoacylation with alanine by alanyl-tRNA synthetase. If aminoglycosides inhibit *trans*-translation in intact bacterial cells, it would interfere with the disposal of miscoded and truncated proteins and would therefore contribute significantly to the antibacterial action of these antibiotics. Evidence for or against such an *in vivo* action is awaited with interest.

5.5.3 Tetracyclines

Several members of this group are illustrated in Figure 5.8. The tetracyclines are broad-spectrum antibiotics with additional activity against rickettsial organisms, mycoplasmas and certain protozoa. The bacteriostatic activity of the tetracyclines results from the direct inhibition of protein biosynthesis. Unlike most other therapeutically useful inhibitors of protein biosynthesis, the tetracyclines inhibit both 70S and 80S ribosomes, although 70S ribosomes are more sensitive. However, the tetracyclines are much more effective against protein synthesis in bacterial cells than against mammalian cells, largely because of the ability of sensitive bacteria to concentrate tetracyclines within their cytoplasm (Chapter 7).

Studies of the effects of the tetracyclines on the tRNA–ribosome interaction show that they inhibit the binding of aminoacyl-tRNA to the A site on the ribosome but have little effect on binding to the P site except at high drug concentrations. The binding of fMet-$tRNA_F$ to the ribosome is about one-tenth as sensitive to tetracycline as the binding of other aminoacyl-tRNAs, since fMet-$tRNA_F$ binds to the P site rather than to the A site. The tetracyclines do not directly inhibit formation of the peptide bond or the translocation step except at high concentrations. They have no effect on the hydrolysis of GTP to GDP required for the functional binding of aminoacyl-tRNA to the A site. Possi-

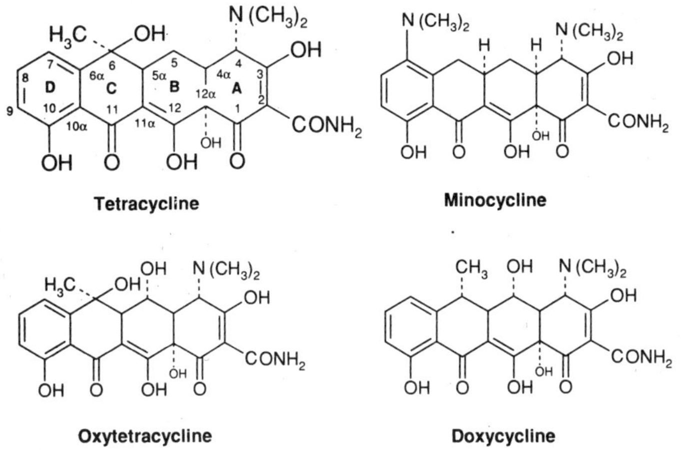

FIGURE 5.8 Four major tetracycline broad-spectrum antibiotics. Minocycline is effective against some bacteria that are resistant to the other drugs.

bly the tetracyclines uncouple GTP hydrolysis from the binding reaction. Tetracyclines also inhibit peptide chain termination and release by blocking the binding of the release factors at stop codons in the A site. However, it is unlikely that the effects on termination and release contribute significantly to the antibacterial action of tetracyclines.

There is a single, moderately high-affinity binding site for tetracycline on the 30S subunit, with a K_d of approximately 1 μM. There are also many low-affinity sites which are not considered essential to the inhibitory action of the drug. Photoaffinity labelling studies with a photoreactive derivative of tetracycline revealed extensive labelling of protein S7, which is located near the region of contact between the two ribosomal subunits. Footprinting studies apparently implicated several residues of 16S rRNA in the binding of tetracyclines. However, doubts regarding the relevance of the footprinting data were raised by the observation that one of the residues, A-892, protected by tetracycline from alkylation by dimethyl sulfate, was not similarly protected by minocycline and doxycycline, which also inhibit ribosomal function in the same way as tetracycline. Furthermore, the multiple binding interactions of tetracyclines with the 30S subunit cause conformational changes in 16S rRNA which may affect the susceptibility of nucleic acid bases to alklylation and the S37 protein to photoaffinity labelling.

Fortunately, X-ray crystallography of tetracycline bound to the 30S subunit has clarified the nature of the binding interaction. Although six tetracycline binding sites were identified, the most important site lies in a pocket formed by the head of the 30S subunit at the A site. The tetracycline molecule interacts with the sugar backbone of helix-34 through a magnesium ion. At this site the antibiotic sterically interferes with the location of the anticodon loop of aminoacyl-tRNA and prevents its binding to the mRNA codon at the A site. The crystallographic data therefore agree well with the classic biochemical results on the action of tetracycline on ribosomal function. The relevance of the other five tetracycline binding sites identified by crystallography to the action of tetracycline is not clear, but they may contribute synergistically to the overall suppression of bacterial protein synthesis.

The features of the tetracycline molecule required for antibacterial activity have been worked out in some detail. Because the permeation mechanism for the entry of tetracyclines into bacterial cells (Chapter 7) may have its own structural requirements, it should not be assumed that the structural features of the tetracycline molecule required for antibacterial activity are the same as those for the inhibition of ribosomal function. It is possible that some tetracycline derivatives which inhibit ribosomes may lack antibacterial activity because of a failure to achieve an inhibitory concentration within the bacterial cell.

The more limited investigations of the structural requirements for the inhibition of protein synthesis on isolated ribosomes reveal several modifications in structure (Figure 5.9) that significantly affect inhibitory activity:

1. Chlorination of the 7 position significantly increases inhibitory activity.
2. Epimerization of the 4-dimethylamino group significantly decreases activity.
3. Both 4α,12α-anhydro- and 5α,6-anhydrotetracyclines are much less active than tetracycline.
4. Ring opening of the tetracycline nucleus to give the iso derivatives and the α and β isomers of apo-oxytetracycline destroys activity.
5. Replacement of the amidic function at C_2 with a nitrile group results in a marked loss of potency.

It has long been suspected that the ability of the tetracyclines to chelate polyvalent cations like Mg^{2+} is relevant to the inhibition of protein biosynthesis. The crystallographic evidence described here now gives strong support for this view. Free Mg^{2+} ions in the bacterial cytoplasm may also complex with tetracycline and limit its ability to interact with ribosomal Mg^{2+}. The 11, 12β-diketone system, the enol (positions 1 and 3) and the carboxamide at position 12 have all been implicated as possible chelation sites. The glycyltetracycline tigicycline (Figure 5.10) is also an effective chelating agent. Another suggestion, based on circular dichroism studies on the 7-chlortetracycline complexes with Ca^{2+} and Mg^{2+}, is that chelation requires the bending of ring A back towards rings B and C so that the oxygen atoms at positions 11 and 12 together with those at positions 2 (amidic oxygen) and 3 form a coordination site into which the metal atom fits.

5.6 Inhibitors of peptide bond formation and translocation

5.6.1 Chloramphenicol

Chloramphenicol (Figure 5.11) is a naturally occurring antibiotic that is now entirely produced by chemical

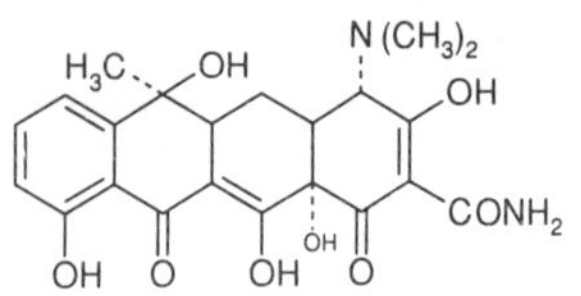

4α, 12α-Anhydrotetracycline

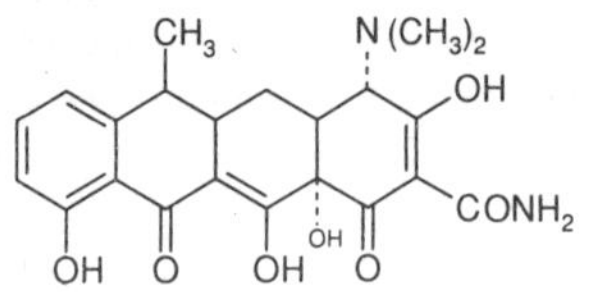

5α, 6-Anhydrotetracycline

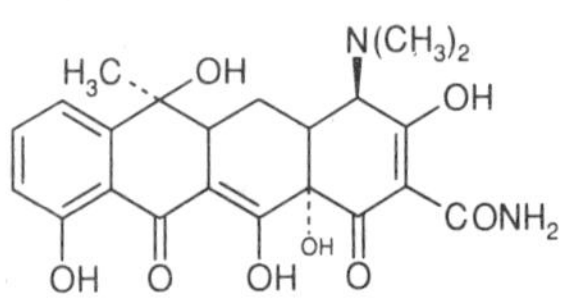

4-Epitetracycline

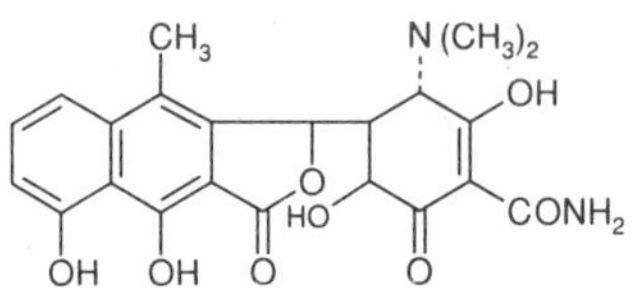

Apo-oxytetracycline

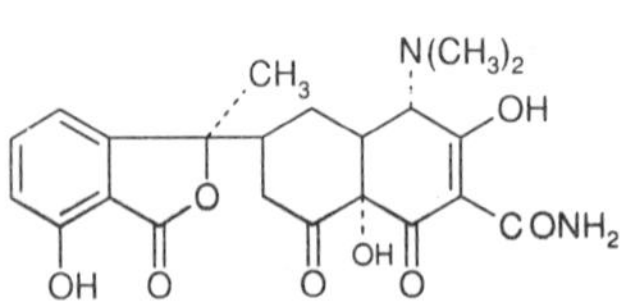

Isotetracycline

FIGURE 5.9 Tetracycline derivatives with greatly reduced antibiotic activity.

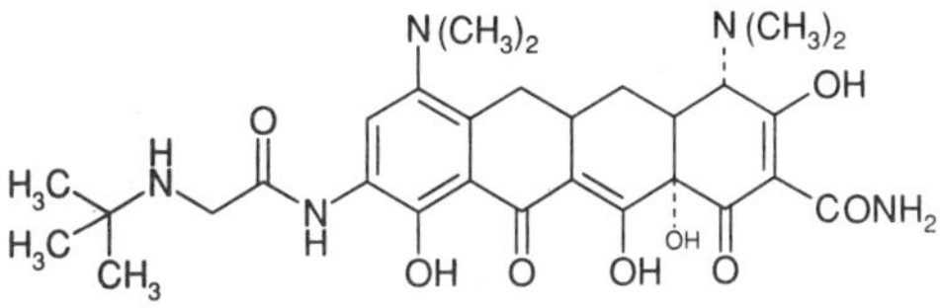

FIGURE 5.10 Tigicycline is a novel tetracycline introduced to combat bacteria resistant to earlier tetracyclines.

synthesis. It was one of the first broad-spectrum antibiotics to be discovered and has excellent bacteriostatic activity against both Gram-positive and Gram-negative cocci and bacilli, as well as rickettsias, mycoplasmas and *Chlamydia*. Unfortunately, its ever-widening use in some parts of the world without effective medical controls revealed serious side effects associated with the bone marrow in some patients. A concentration of 25–30 μg of chloramphenicol for ml^{-1} of blood maintained for 1–2 weeks leads to an accumulation of nucleated erythrocytes in the marrow, indicating an interference with the normal red cell maturation process. Normal erythropoiesis usually resumes after withdrawal of the drug, but very occasionally, i.e. not more than 1 in 20,000 cases, a more serious defect develops in the marrow which leads irreversibly to the loss of both white and red cell precursors. The biochemical basis for chloramphenicol-induced fatal aplastic anemia has not been established. However, an action of the drug on mitochondrial ribosomes, which more closely resemble 70S than 80S ribosomes, with possible effects on the mitochondrial function of key stem cells in the marrow, cannot be ruled out. The very low incidence of the irreversible form of chloramphenicol toxicity indicates a special sensitivity in those few individuals who succumb to it. Chloramphenicol therapy is therefore now restricted to serious infections for which there is no effective alternative. These include

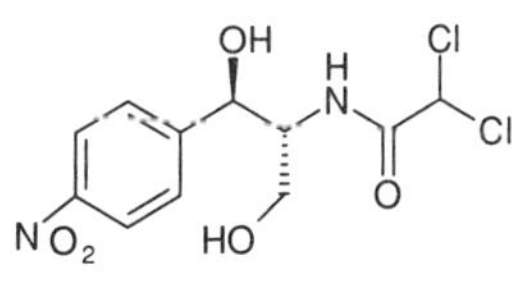

FIGURE 5.11 Chloramphenicol. The active form is the D-*threo* stereoisomer.

typhoid and meningitis caused by bacteria resistant to β-lactam antibiotics, in cases where the patient is allergic to these drugs or where use is restricted to topical application, for example in superficial infections of the eye.

The bacteriostatic action of chloramphenicol is due to a specific inhibition of protein biosynthesis on 70S ribosomes; it is completely inactive against 80S ribosomes. Studies with radioactively labelled chloramphenicol show that it binds exclusively to the 50S subunit to a maximum extent of one molecule per subunit. The binding is completely reversible. Structurally unrelated antibiotics such as erythromycin and lincomycin, that also interfere with the function of the 50S subunit, compete with chloramphenicol for the binding region, whereas inhibitors of the 30S subunit function, such as the aminoglycosides and tetracyclines, do not. Biochemical evidence strongly indicates that chloramphenicol blocks peptide bond formation by inhibiting the peptidyl transferase activity of the 50S subunit. Thus chloramphenicol inhibits the puromycin-dependent release of nascent peptides from 70S ribosomes and also inhibits the 50S subunit-catalyzed puromycin fragment reaction.

Earlier studies concentrated on the identification of proteins thought to be involved in the binding of chloramphenicol to the 50S subunit. The technique of affinity immune electron microscopy suggested that the peptidyl transferase region of the ribosome, containing the proteins L15, L18 and L27, contributed to the binding of chloramphenicol. Because the vast majority of known enzymes are proteins, it is understandable that the early work on the action of chloramphenicol sought to identify a protein or proteins associated with peptidyl transferase activity that would provide the target site for the antibiotic. However, the discovery that domain V of 23S rRNA catalyzes peptide bond formation led to a reassessment of the nature of the target for chloramphenicol. Although the results of footprinting studies suggested that several nucleotides of 23S rRNA were involved in the interaction of chloramphenicol with its binding site, once again it remained for X-ray crystallography to reveal more precise details of the binding site. *Deinococcus radiodurans* 50S subunits complexed with chloramphenicol confirmed that the drug is bound to a hydrophobic crevice in the peptidyl transferase region of the A site. The X-ray data provide evidence that the two oxygen atoms of the nitro group of chloramphenicol, the 1-hydroxyl, 3-hydroxyl and 4′-carboxyl groups all have the potential to hydrogen bond with several specific nucleotide bases in a loop of domain V of 23S rRNA. A Mg^{2+} ion may also be involved in the bonding of the 3-hydroxyl with adjacent nucleotide bases. The antibiotic is thus exquisitely placed to block the correct positioning of the incoming aminoacyl tRNA and so inhibit the formation of the next peptide bond. An analogous study with 50S subunits from the archaeobacterium *Haloarcula marismortui* placed the binding site away from the peptidyl transferase center in another hydrophobic crevice at the entrance to the polypeptide exit tunnel. Because pathogenic bacteria and *Deinococcus radiodurans* belong to the eubacteria, it seems likely that the binding data from *Deinococcus* may be more relevant to the therapeutic action of chloamphenicol than the *Haloarcula* data, although it can be argued that binding to both sites may important in the antibacterial action of the drug.

5.6.2 Clindamycin

This is a semisynthetic member of the naturally occurring lincosamide antibiotics. Clindamycin (Figure 5.12) is clinically useful against staphylococcal and streptoccocal infections and against anaerobic pathogens such as *Bacteroides* spp. It is poorly active against Gram-negative bacteria. Like chloramphenicol, clindamycin inhibits the peptidyl transferase activity of the 50S subunit. However, whereas chloramphenicol interferes with the correct positioning of aminoacyl tRNAs at the active center, clindamycin disrupts substrate binding and hinders the path of the growing peptide chain. The binding site of clindamycin partially overlaps with that of chloramphenicol at the A site, but the drug also interacts with the P site. Crystallographic data show that clindamycin binds to a specific loop of domain V of 23S rRNA. There is no evidence of any interactions with ribosomal proteins. The three hydroxyl groups of the sugar moiety of clindamycin are all positioned in the crystal structure to form hydrogen bonds with specific nucleotide bases of the 23S rRNA. The ability of clindamycin to interact with the A site is probably due to the proline moiety of the drug, which is seen to overlap with the phenyl ring of chloramphenicol at the A site. The carbon atom of the methyl group at the terminal of the sugar side chain residue is positioned to interact with a nucleotide base at the P site, possibly involving van der Waals and hydrophobic forces. The sulfur atom of clindamycin may also interact with the P site.

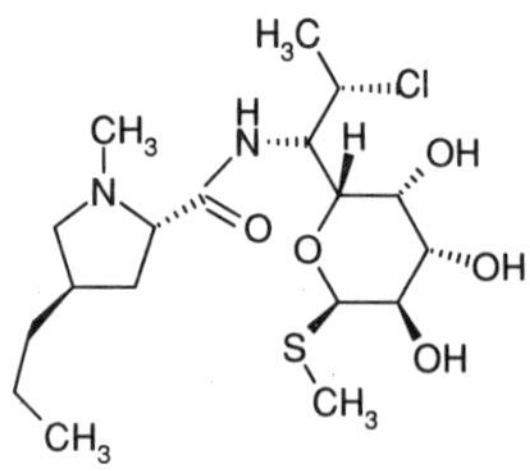

FIGURE 5.12 Clindamycin is a narrow-spectrum antibacterial drug effective against Gram-positive bacteria .

5.6.3 Macrolides and ketolides

The original macrolide antibiotic, erythromycin (Figure 5.13), is a complex, naturally occurring compound effective against many Gram-positive bacteria, mycoplasmas and chlamydia, but less active against Gram-negative pathogens. Semisynthetic derivatives, e.g. azithromycin, brought additional benefits, including useful activity against *Haemophilus influenzae*, an important Gram-negative pathogen of the upper respiratory tract. The most important recent introductions have been the ketolides like telithromycin (Figure 5.13) in which the cladinose sugar moiety at position C-3 is replaced with a keto group. A major advantage of the ketolides is that they retain activity against some bacteria that have become resistant to the older macrolides.

Despite the various chemical differences among macrolides and ketolides, the drugs share a common mode of action; they inhibit the function of the 50S subunit by blocking the access of the growing peptide chain to the entrance of the polypeptide exit tunnel. Thus, although they do not directly inhibit peptidyl transferase activity, macrolides and ketolides effec-

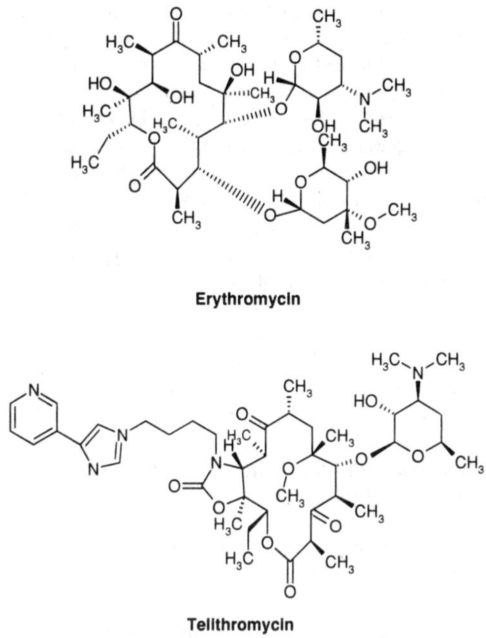

FIGURE 5.13 Erythromycin, a 'medium-spectrum' antibacterial drug of the macrolide family and telithromycin, a semisynthetically modifed macrolide antibiotic, are representative of a new series of ketolides.

tively suppress protein biosynthesis in bacteria. They are without effect on the 80S ribosomes of mammalian cells. Details of the actual binding site for erythromycin on the 50S subunit emerging from X-ray crystallography data pinpoint the potential for multiple hydrogen bond interactions between the desosamine ring at the 5 position of the macrolactone moiety of the antibiotic with nucleotides of domain V of 23S rRNA. This domain is a major component of the exit tunnel. Van der Waals forces may also contribute to the stability of the binding interaction. A commonly encountered form of bacterial resistance to erythromycin depends on *N*-dimethylation by the inducible enzyme *N*-methyl transferase at the N-6 position of a specific adenine residue of 23S rRNA: A-2058 in *Escherichia coli*, A-2086 in *Bacillus stearothermophilus* and A-2058 in *Bacillus subtilis* (see Chapter 9 for further details of the *N*-methyl transferase system). The crystal structure of the 50S subunit shows clearly that methyl groups in the indicated position prevent hydrogen bonding between A-2058 (of *Deinococcus radiodurans*) and the 2′ hydroxyl group of the desosamine ring of erythromycin and other macrolides. The methyl groups would very likely also sterically hinder access of the antibiotic to its binding site. This illustrates a pleasing concordance between X-ray studies and classic biochemical work on antibiotic resistance. It is interesting that humans and all other mammals have guanine instead of adenine at position 2058 in 23S rRNA, which effectively prevents macrolides and ketolides from binding to 80S ribosomes.

The crystal structure of telithromycin, bound to the *Deinococcus radiodurans* 50S subunit, shows that the antibiotic also blocks the ribosomal exit tunnel. The interactions involve domains II and V of 23S rRNA and provide tighter binding than macrolide antibiotics—an observation borne out by direct binding studies. The tighter binding of telithromycin to ribosomes may account for its improved activity against macrolide-resistant bacteria harbouring the *N*-dimethylase enzyme.

5.6.4 Streptogramins

The streptogramin antibiotics were discovered more than four decades ago but until recently were of more academic than practical interest. The growing menace of bacterial resistance to antibiotics, however, raised the possibility that some members of this large group of complex compounds could be clinically useful. Streptogramins are of two structurally distinct types which are produced as mixtures by certain *Streptomyces*. Type A streptogramins are polyunsaturated cyclic lactones that resemble the macrolides. Type B streptogramins are cyclic hexadepsipeptides. More recently, semisynthetic modifications of the naturally occurring streptogramins have yielded significant improvements in clinically important properties such as water solubility. Dalfopristin and quinupristin (Figure 5.14) are examples of semisynthetic derivatives of type A and type B compounds, respectively. A combination of dalfopristin and quinupristin is proving valuable in the treatment of serious and potentially life-threatening infections caused by methicillin-resistant

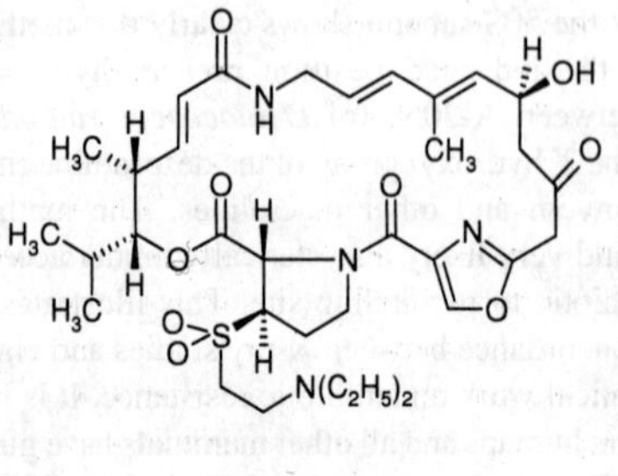

Dalfopristin

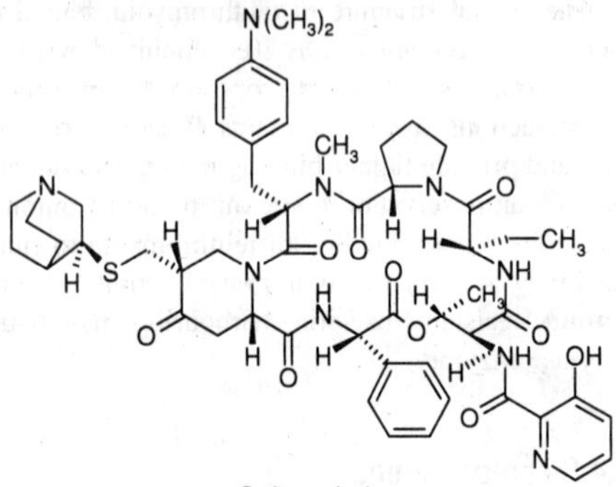

Quinupristin

FIGURE 5.14 Dalfopristin and quinupristin are streptogramin antibiotics that, when combined, cause irreversible inhibition of bacterial protein biosynthesis.

Staphylococcus aureus and vancomycin-resistant enterococci (VRE).

Remarkably, the two types of antibiotic act synergistically to give irreversible inhibition of protein biosynthesis and consequently a bactericidal effect. The synergism arises from the distinct actions of the different streptogramins on ribosomal function. Type A compounds block the binding of aminoacyl-tRNA and peptidyl-tRNA to the A and P sites, respectively. Type B compounds hinder the interaction of peptidyl-tRNA only with the P site. Some form of conformational change in ribosomal structure is induced by type A compounds and enhances the affinity of the 50S subunit for type B streptogramins. X-Ray analysis of the 50S subunit from *Haloarcula marismortui* cocrystallized with the streptogramin A antibiotic, virginiamycin M, revealed that the conformational change in the subunit induced by this antibiotic results from its binding to portions of both the A and P sites. It will be of interest to see whether this mode of binding of type A streptogramins is replicated in ribosomes belonging to bacterial species other than archaeobacteria like *Haloarcula marismortui*. *N*-Dimethylation of A-2058 in domain V of 23S rRNA or its replacement by guanine or uridine leads to resistance to the type B streptogramins, indicating a role for this adenine residue in the binding interaction. It is interesting that modification or replacement of A-2058 causes resistance to structurally diverse compounds, including macrolides, lincosamides and type B streptogramins.

5.6.5 Oxazolidinones: novel synthetic inhibitors of peptidyl transferase

The antibacterial activity of oxazolidinones was discovered more than 20 years ago, but toxicity problems with the early compounds prevented their further development. Fortunately, a re-examination of the series in the early 1990s led to the synthesis of much less toxic compounds and the discovery of linezolid (Figure 5.15). Linezolid has excellent activity against dangerous Gram-positive infections caused by MRSA and VRE. Because linezolid is representative of the first chemically novel series of antibacterial drugs to emerge in several decades, there is now intense interest in this area of chemistry. Linezolid and most other oxazolidinones developed so far lack useful activity against Gram-negative pathogens, probably because the drug efflux pumps in these bacteria are very effective in clearing the compounds from bacterial cyto-

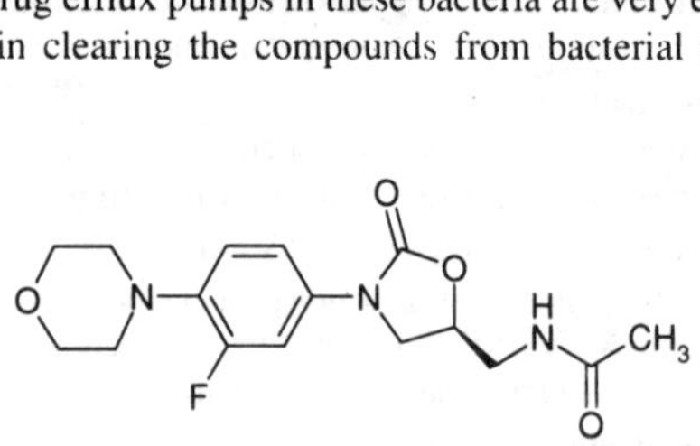

FIGURE 5.15 Linezolid was the first clinically active oxazolidinone antibiotic. It is effective against Gram-positive bacteria.

plasm (see Chapter 7). Much attention is therefore currently focused on synthesizing novel oxazolidinones with good activity against Gram-negative pathogens such as *Haemophilus influenzae*. The Enterobacteriacae may provide an even greater challenge.

Following the revival of interest in oxazolidinones, it was soon discovered that the basis of their antibacterial action is the inhibition of protein biosynthesis. Locating the precise site of action has proven more difficult. It became clear that oxazolidinones bind exclusively to the bacterial 50S subunit and prevent complexation with its 30S partner, mRNA, initiation factors and fMet-tRNA$_f$. This appeared to place the inhibitory mechanism in a separate category from the existing inhibitors of protein biosynthesis which act at points beyond the formation of the preinitiation complex. However, later experiments suggested that oxazolidinones inhibit the binding of fMet-tRNA$_F$ to the peptidyl transferase at the P site of the 50S subunit. A study of the effects of mutations affecting nucleotides of 23S rRNA which confer resistance to oxazolidinones found that the nucleotides are restricted to a confined space between the A and P sites in the three-dimensional structure of the 50S subunit. This space partially overlaps the P site. Another recent investigation used a photoactivated, radioiodine-labelled oxazolidinone derivative to identify drug-binding sites in ribosomes actively engaged in protein synthesis (in contrast to isolated purified ribosomes). The compound cross-linked with A-2602, which is known to be a critical base in the peptidyl transferase active center. Protein L27, believed to be closely associated with, although not functionally involved with peptidyl transferase, was also labelled. Even allowing for possible artefacts arising in photoaffinity labelling experiments, i.e. spurious labelling of irrelevant 'bystander' molecules, the weight of evidence from both the affinity labelling and mutational studies points to the peptidyl transferase center as the primary site of oxazolidinone action. Interference with the correct alignment of fMet-tRNA$_F$ at the P site could explain the disruption of the formation of the preinitiation complex.

The surprisingly low binding affinity of oxazolidinones for isolated, purified 50S subunits has so far precluded an X-ray crystallographic analysis of the drug–ribosome interaction. However, the considerable potential for the clinical application of oxazolidinone antibiotics will no doubt continue to focus attention on the precise details of their mechanism of action.

5.6.6 Fusidic acid

Fusidic acid belongs to a group of steroidal antibiotics (Figure 5.16) which inhibit the growth of Gram-positive but not Gram-negative bacteria. The inactivity of fusidic acid against Gram-negative bacteria may be due to inadequate penetration into the bacterial cytoplasm, since the drug inhibits protein synthesis on isolated ribosomes from Gram-negative bacteria. Alternatively, fusidic acid may be rapidly extruded from Gram-negative cells by one or more of the drug-efflux pumps (Chapter 7). Fusidic acid is used topically in some countries to treat Gram-positive infections of the skin that cause impetigo. Occasionally the drug may be given systemically to combat potentially life-threatening infections caused by MRSA and VRE. However, fusidic acid is not available in the United States.

The addition of fusidic acid to 70S ribosomes *in vitro* prevents the translocation of peptidyl-tRNA from the A site to the P site and eventually suppresses the EF-G-dependent hydrolysis of GTP. The inhibition is overcome by the addition of excess EF-G. The resistance of some strains of bacteria to fusidic acid appears to be associated with a change in EF-G, since the factor prepared from resistant cells normally catalyzes translocation in the presence of the antibiotic. All this points to factor EF-G as the target protein for fusidic acid. Fusidic acid-hypersensitive mutant forms of EF-G have been used in an effort to locate the binding site

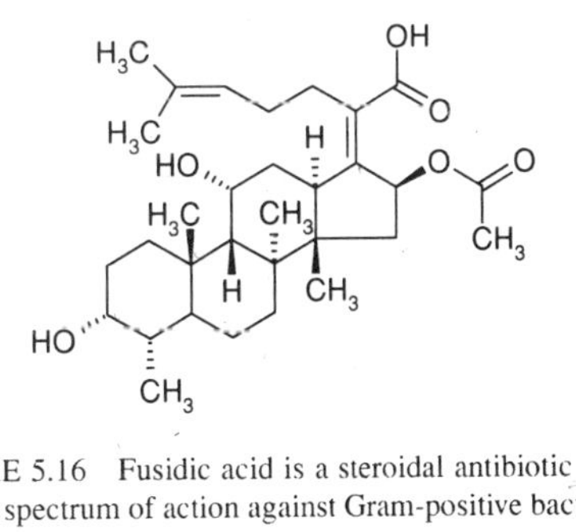

FIGURE 5.16 Fusidic acid is a steroidal antibiotic with a narrow spectrum of action against Gram-positive bacteria.

for the drug. However, it seems unlikely that the amino acids involved in these mutations could all make contacts with the drug. Instead it is believed that the mutations probably modify the conformation of EF-G and enhance its affinity for fusidic acid. A current model for the action of fusidic acid proposes that the initial EF-G–ribosome interaction triggers a burst of GTP hydrolysis which either uncovers or even generates the binding site for fusidic acid on EF-G. A stable complex is formed of EF-G-GDP-fusidic acid and the ribosome which is unable to release EF-G for a further round of translocation and GTP hydrolysis. Fusidic acid also inhibits protein synthesis on 80S ribosomes in a similar manner by stabilizing the EF-2-GDP-ribosome complex (EF-2 corresponds to the prokaryotic EF-G). The lack of toxicity of fusidic acid against mammalian cells is probably because the drug does not achieve a high enough intracellular concentration in mammalian cells to form the stabilized complex.

5.6.7 Cycloheximide

Occasionally referred to as actidione, cycloheximide (Figure 5.17) is toxic to a wide range of eukaryotic cells, including protozoa, yeasts, fungi and mammalian cells, and its lack of selectivity precludes any clinical use. It is included here because of its unusual specificity of action against 80S ribosomes, with no effect on 70S ribosomes. Cycloheximide is therefore mostly used as an experimental tool to inhibit protein synthesis in eukaryotic cells, and occasionally to exclude fungi from bacterial cultures.

There are considerable variations in the sensitivity of 80S ribosomes from different species of yeasts to cycloheximide, which have been exploited to explore its site of action. For example, ribosomes from *Saccharomyces cerevisiae* are strongly inhibited by cycloheximide, while those from *Saccharomyces fragilis* and *Kluyveromyces lactis* are resistant. Cross-over experiments with the 60S and 40S subunits from *Saccharomyces cerevisiae* and *Saccharomyces fragilis* showed that sensitivity to cycloheximide resides in the 60S subunit. Analysis of the proteins of the 60S subunit from *Kluyveromyces lactis* revealed that protein L41 differs from the corresponding protein in *Saccharomyces cerevisiae,* suggesting that L41 may be involved in the interaction of the antibiotic with susceptible ribosomes. A recent study found that L41 contributes to the efficiency of protein synthesis in *Saccharomyces* cerevisiae but is not essential. Mutants in which L41 was deleted had an increased rate of elongation, indicating that L41 in some way normally regulates the rate of elongation. The L41-deficient mutants were more resistant to cycloheximide, suggesting that the compound inhibits the translocation of peptidyl-tRNA from the A to the P site. Other results show that two other ribosomal proteins, L42 and L28, also contribute in some way to the interaction of cycloheximide with 80S ribosomes. Footprinting studies reveal that cycloheximide protects two guanine residues lying within a loop region of 28S rRNA of the larger ribosomal subunit which is associated with the hydrolysis of GTP and the ribosomal interaction with elongation factor EF-2. This is consistent with the inhibition by cycloheximide of the translocation step, and it seems likely that the antibiotic hinders the function of EF-2. The involvement of proteins L41, L42 and L28 in the action of cycloheximide may be linked to the role of these proteins in maintaining the conformation of 28S rRNA so as to afford a nucleic acid binding site for the antibiotic.

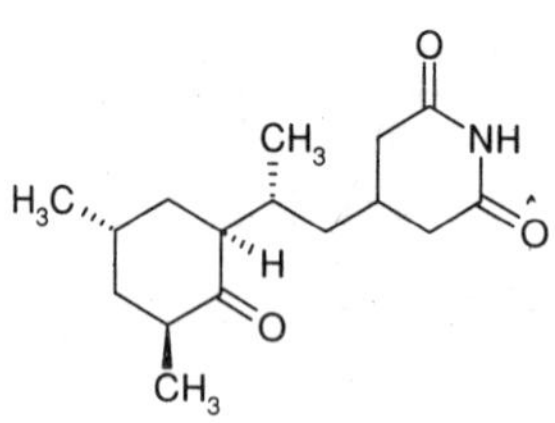

FIGURE 5.17 Cycloheximide is an unusually specific inhibitor of 80S ribosomes.

Further reading

Auerbach, T. *et al.* (2002). Antibiotics targeting ribosomes: crystallographic studies. *Curr. Drug Targets Infect. Disord.* **2**, 169.

Ban, N. *et al.* (2000). The complete atomic structure of the large ribosomal subunit at 2.4Å resolution. *Science* **289**, 905.

Bashan, A. *et al.*, (2003). Structural basis of the ribosomal machinery for peptide bond formation, translocation and nascent chain progression. *Molec. Cell.* **11**, 91.

Berisio, R. *et al.* (2003). Structural insight into the antibiotic action of telithromycin against resistant mutants. *J. Bact.* **185**, 4276.

Bobkova, E. V. *et al.* (2003). Catalytic properties of mutant 23S ribosomes resistant to oxazolidinones. *J. Biol. Chem.* **278**, 9802.

Bonfiglio, G. and Furneri, P. M. (2001). Novel streptogramin antibiotics. *Exp. Opin. Investig. Drugs.* **10**, 185.

Carter, A. P. *et al.* (2000). Functional insights from the structure of the 30S ribosomal subunit and its interactions with antibiotics. *Nature* **407**, 340.

Chopra, I. and Roberts, M. (2001). Tetracycline antibiotics: mode of action, applications, molecular biology and epidemiology of bacterial resistance. *Microbiol. Molec. Biol. Reviews* **65**, 232.

Colcat, J. R. *et al.* (2003). Cross-linking in the living cell locates the site of action of oxazolidinone antibiotics. *J. Biol. Chem.* **278**, 21972.

Corvaisier, S., Bordeau, V. and Felden, B. (2003). Inhibition of transfer RNA aminoacylation and *trans*-translation by aminoglycoside antibiotics. *J. Biol. Chem.* **278**, 14788.

Dresios, J. *et al.* (2003). A dispensable yeast ribosomal protein optimizes peptidyl transferase activity and affects translocation. *J. Biol. Chem.* **278**, 3314.

Hansen, J. L. *et al.* (2003). Structure of five antibiotics bound at the peptidyl transferase center of the large ribosomal subunit. *J. Mol.Biol.* **330**, 1061.

Livermore, D. (2003). Linezolid *in vitro* : mechanisms and antibacterial spectrum. *J. Antimicrob. Chemother.* **51**, Suppl. S2, ii9.

Martemyanov, K. A. *et al.* (2001). Mutations in the G-domain of elongation factor G from *Thermus thermophilus* affect both its interaction with GTP and fusidic acid. *J. Biol. Chem.* **276**, 28774.

Nakama, T., Nurecki, O. and Yokoyama, S. (2001). Structural basis for the recognition of isoleucyl-adenylate and an antibiotic mupirocin by isoleucyl-tRNA synthetase. *J. Biol. Chem.* **276**, 47387.

Ogle, J. M. *et al.* (2003) Insights into the decoding mechanism from recent ribosome structures. *Trends Biochem. Sci.* **28**, 259.

Poehlsgaard, J. and Douthwaite, S. (2002). The macrolide binding site on the bacterial ribosome. *Curr. Drug Targets Infect. Disord.* **2**, 67.

Schluenzen, F. *et al.* (2001). Structural basis for the interaction of antibiotics with the peptidyl transferase center in eubacteria. *Nature* **413**, 814.

Vannuffel, P. and Cocito, C. (1996). Mechanisms of action of streptogramins and macrolides. *Drugs* **51**, Suppl. 1, 20.

Zhanel, G. C. *et al.* (2002). The ketolides. *Drugs* **62**, 1771.

Chapter six

Antimicrobial drugs with other modes of action

The antimicrobial agents described so far have been arranged according to their primary effects on cellular metabolism and biosynthesis. This approach accounts for many of the most important drugs in current antimicrobial therapy. There are, however, several valuable drugs whose various modes of action fall outside those most commonly encountered. In this chapter we describe important examples of compounds with unusual actions. The drugs are arranged according to their therapeutic targets—antibacterial, antifungal, antiviral and antiprotozoal—although some of the antibacterial drugs also have useful activity against certain protozoal pathogens.

6.1 Nitroheterocyclic antimicrobial drugs

The discovery of the naturally occurring nitroaromatic antibiotic chloramphenicol (Figure. 5.11) raised the possibility that other organic nitro compounds would have antimicrobial activity. The five compounds shown in Figure 6.1 are perhaps the most useful and commonly used antimicrobial drugs among the many nitro compounds that have been screened. Some other nitro compounds have important applications in the eradication of parasitic nematode worms, but these fall outside the scope of this book.

Metronidazole is valuable in the treatment of infections caused by strictly anaerobic bacterial pathogens such as *Bacteroides fragilis* and *Clostridium difficile*, and against *Helicobacter pylori*, a bacterium causally linked to peptic ulcer disease and gastritis which inhabits the stomach in an acid environment of low oxygen tension. The spectrum of metronidazole extends to several protozoal parasites, including *Giardia lamblia, Entamoeba histolytica* and *Trichomonas vaginalis*. Nitrofurazone is a broad-spectrum antibacterial agent, useful as a topical treatment for infected burns and skin grafts. Nitrofurantoin is sometimes described as a urinary antiseptic, owing to its value in clearing infections of the urinary tract caused by Gram-negative pathogens such as *Pseudomonas aeruginosa, Serratia marscescens* and *Proteus mirabilis*. Furazolidone also has antibacterial activity, but its principal use is against certain protozoal infections, including *Giardia lamblia*. Although benznidazole is included in Figure 6.1, its only application is in the treatment of South American trypanosomiasis, or Chagas' disease (see later discussion).

The nitroheterocyclic drugs are all subject to reduction to bioactive molecules or radicals by cellular enzymes of the target pathogens. Their spectrum of action is mainly a function of the redox potentials of the consituent nitro groups. Nitrofurazone, nitrofurantoin and furazolidone have relatively high redox potentials of between –250 and –270 mV, whereas metronidazole has a much lower potential of –480 mV. The first three drugs can be reductively activated by a wide range of enzymes such as the NADP(H)-dependent nitro reductases. Owing to the much lower redox potential of metronidazole, the drug can only be activated by the pyruvate-ferredoxin oxido reductases and hydroge-

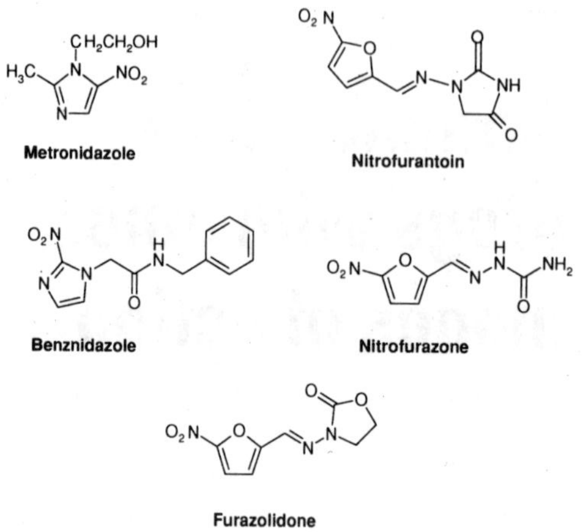

FIGURE 6.1 Nitroheterocyclic drugs with activity against anaerobic bacteria and some protozoan parasites.

nases of anaerobic bacteria and protozoa such as *Trichomonas vaginalis*. However, in all cases it is considered that the nitro groups are reduced to a short-lived nitro radical anion:

$$R\text{—}NO_2 \leftrightarrow R\text{—}NO_2^{\bullet -}$$

or other short-lived products, including chemically reactive hydroxylamine derivatives. The bioactive products of the reductive process of metronidazole attack DNA, causing single- and double-stranded breaks and base mutations. Although the precise nature of the damage to DNA is not clear, there is a preferential attack of the reactive metabolite on thymidine residues and other pyrimidines. Bacteria with mutations that adversely affect excision repair and DNA recombination are more sensitive to metronidazole. Damage to DNA is believed to be the cause of cell death and lysis in bacteria and protozoal parasites. The situation appears to be rather different with the three other nitroheterocylic drugs. Their bioactive products do not cause fragmentation of DNA, although they are weakly mutagenic. It is possible that the reductive activation of these drugs generates short-lived intermediates that target other essential biochemical processes, although further research is needed to investigate this possibility.

The model presented here accounts for the activity of metronidazole against anaerobic bacteria. The situation is less clear in the case of organisms which survive under conditions of low, rather than zero, oxygen tension, such as *Helicobacter pylori*. One suggestion is that in the presence of oxygen, the nitro radical anion is converted back to metronidazole and superoxide:

$$R\text{—}NO_2^{\bullet -} \leftrightarrow R\text{—}NO_2 + O_2^{\bullet -}$$

The superoxide generated in this process is then converted by the enzyme superoxide dismutase to hydrogen peroxide and molecular oxygen. In the presence of a transition metal, either iron or copper, a series of reactions, known collectively as the Haber-Weiss reaction, give rise to the highly reactive hydroxyl radical, $OH^{\bullet}$, which also damages DNA, again resulting in cell death. The ability of nitroheterocyclic drugs to damage DNA increases the risk of harmful mutagenic effects in the infected patient.

6.2 A unique antifungal antibiotic—griseofulvin

Fungal infections of the skin and nails ('ringworm') are commonly caused by various species of *Trichophyton*. Such infections usually respond well to topically applied drugs of the azole class (Chapter 3). However, where the infection is widespread, oral treatment with azoles or terbinafine may be necessary. In cases where the fungus is both widespread and resistant to these drugs, the physician may turn to griseofulvin (Figure 6.2). This is an antibiotic produced by *Penicillium griseofulvum* which causes the tips of fungal hyphae to become curled and suppresses further growth. At the molecular level, it binds to the intracellular protein tubulin and possibly to the accessory proteins involved in the polymerization of tubulin to form microtubules (the microtubule-associated proteins, MAPs). Microtubules participate in the movements of subcellular organelles and in the separation of daughter chromosomes during mitosis in eukaryotic cells. A characteristic property of microtubules is their rapid assembly and disassembly which is due to the reversible polymerization of the constituent tubulin. This dynamic in-

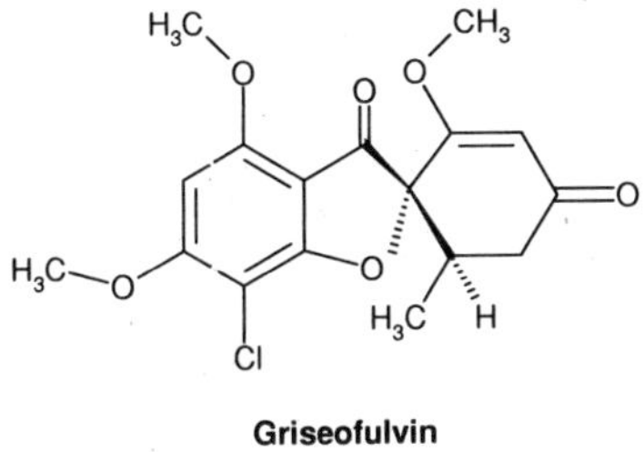

FIGURE 6.2 Griseofulvin, a unique antifungal antibiotic that disrupts the function of microtubules.

stability, as it is called, is crucial to the mitotic process. The binding of griseofulvin to tubulin and the MAPs in some way hinders the assembly and disassembly of the microtubules and thereby disrupts cell proliferation.

The selectivity of griseofulvin for dermatophytes is not fully understood since the antibiotic also binds to mammalian tubulin. Nevertheless, the many differences in the amino acid sequences of, for example, the tubulin of the yeast *Saccharomyces pombe* and that of mammalian brain tubulin may permit differential binding of griseofulvin to the tubulin of fungal cells under the conditions of antifungal therapy.

Griseofulvin has recently been shown to disrupt the function of tubulin in mitosis in the protozoan parasite *Trichomonas vaginalis*. In view of concern about the potential mutagenicity of metronidazole, griseofulvin might provide an alternative treatment for infections caused by this organism.

6.3 Antiviral agents

6.3.1 Inhibitors of the protease of the human immunodeficiency virus

In Chapter 4 we reviewed inhibitors of HIV reverse transcriptase. While these drugs have provided a major advance in the treatment of AIDS, their long-term efficacy in monotherapy is always threatened by the remarkable ability of HIV to generate drug-resistant mutants. To combat this tendency, inhibitors of reverse transcriptase are usually given in combination with other inhibitors of HIV replication. Outstanding among the latter agents is an expanding group of compounds that inhibit the HIV-specific protease.

The viral translation products formed during the replicative phase of the virus are long polypeptide precursors which are then specifically cleaved to release several mature proteins. In this way the virus generates a number of distinct proteins from a single mRNA molecule. The enzyme responsible for the cleavage process, HIV protease, is itself first formed as a zymogen precursor protein. HIV protease, encoded by the viral *pol* gene, belongs to the aspartyl class of proteases. Enzymes of this mechanistic type are maximally active in the pH range 4.5–6.5 but nevertheless retain significant activity at the cytoplasmic, neutral pH.

As the nature of HIV and its replicative cycle were revealed during the 1980s, the importance of HIV protease as a potential target for antiviral activity became apparent. Considerable attention had already been given to the design of inhibitors of another aspartyl protease, renin, and the peptidic inhibitors of this enzyme provided a good starting point for the development of anti-HIV protease compounds. However, like many of the renin inhibitors, the early HIV protease inhibitors were flawed by extremely low solubility, which resulted in poor uptake and tissue distribution ('bioavailability') after administration. Then in 1989 X-ray crystallography revealed the three-dimensional structure of HIV protease, which greatly facilitated the design of nonpeptidic inhibitors. Like the peptides, the nonpeptidic compounds are also competitive inhibitors of the natural substrate. Unlike the earlier peptidic inhibitors, the best of the nonpeptidic compounds have acceptable solubility and bioavailability. As a result, a range of HIV protease inhibitors is now in clinical use and new compounds continue to be developed. The four examples given in Figure 6.3 are potent selective inhibitors that produce rapid decreases in the number of viruses in the blood and, significantly, an increase in the circulating CD4 lymphocytes that are so critical to effective immune defence against microbial infections.

HIV protease inhibitors are now an established component of the HAART regime (see Chapter 4) in combination with at least two different inhibitors of HIV reverse transcriptase. This therapy can reduce the circulating viral RNA in 90% of patients to levels that are not detectable by the polymerase chain reaction for

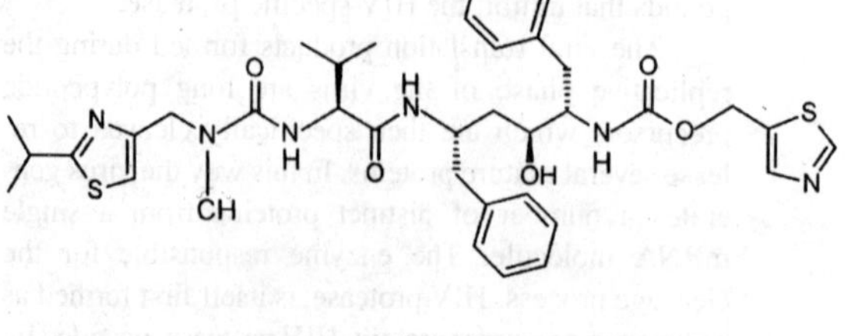

Ritonavir

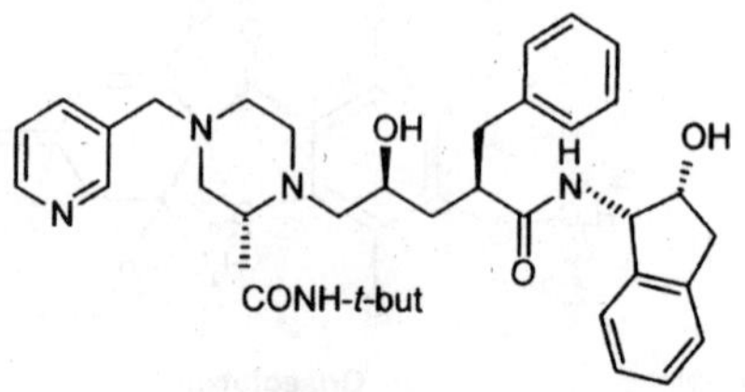

Indinavir

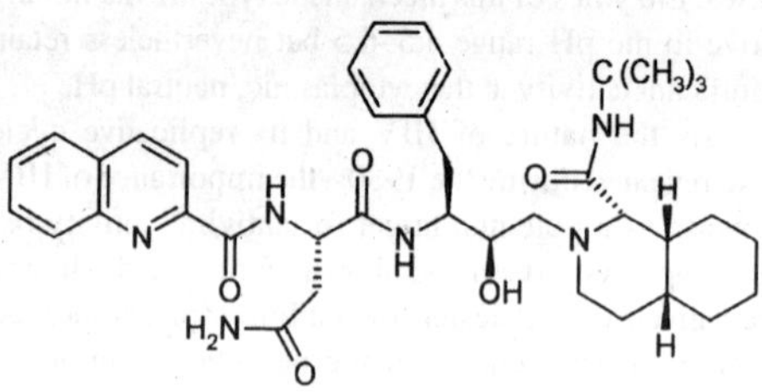

Saquinavir

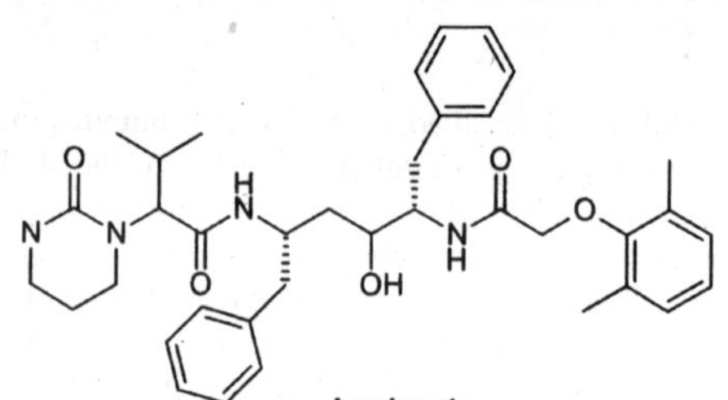

Lopinavir

FIGURE 6.3 Inhibitors of HIV protease used in the therapy of AIDS in combination with other antiretroviral drugs.

up to one year. Ominously, however, replication of the virus may persist in lymph nodes despite sustained drug therapy.

None of the currently available anti-HIV drugs are free of side effects, and these can make it difficult for patients to maintain the strict regime of drug taking that is necessary to sustain the effectiveness of HAART. An unexpected side effect of the protease inhibitors is the suppression of insulin secretion from the pancreatic islet cells in response to a rise in circulating glucose levels. This leads to insulin resistance, hyperlipidemia, and in an unfortunate minority of patients, type II diabetes. The mechanism of this serious problem is being vigorously pursued with a view to designing new inhibitors of HIV protease that are free from effects on the insulin response.

6.3.2 Inhibition of HIV entry into host cells

Because of the remarkable ability of HIV to acquire resistance to virtually any drug it is confronted with, the search for inhibitors of other key stages in the life cycle of the virus continues. HIV gains entry to cells in a three-step process: (a) The viral envelope protein Env binds to the CD4 glycoprotein receptor on the surface of CD4$^+$ T cells, macrophages, dendritic cells and monocytes. (b) The binding of the homotrimeric Env to CD4 induces a conformational change in the Env subunit gp120 that allows concomitant binding to either of the cell surface chemokine receptors, CCR5 and CCR4. (c) Coreceptor binding triggers exposure of a hydrophobic fusion peptide at the N-terminal of the Env subunit gp41, which then inserts into the membrane of the host cell, causing fusion of the viral membrane with the host membrane. This membrane fusion process ensures the release of the viral core protein and the RNA strands into the cytoplasm. Both the binding steps and the membrane fusion process are currently being targeted in drug design programs and several compounds are under clinical evalution in HIV-infected patients, with promising results. The fusion inhibitor enfuvirtide (T20, Fuzeon), which is a synthetic peptide with the sequence Ac-YTSLIHSLIEESQN

QQEKNEQELLELDKWASL-NH_2, binds to a region of gp41 which is exposed transiently after Env binds to CD4 and thereby disrupts the ensuing steps in the fusion–viral entry process.

6.3.3 Antiinfluenza drugs

Influenza is potentially a very dangerous infection; the great influenza pandemic of 1918–1919 is said to have killed some 21 million people worldwide. While the subsequent pandemics have generally not approached the severity of the 1918–1919 outbreak, there is continuing concern that even relatively mild, localized epidemics can cause considerable social and economic dislocation and significant mortality. The main defense against influenza is annual immunization with inactivated, killed virus vaccines, constantly modified to take account of the propensity of the influenza virus for mutation and recombination. Despite the reasonable success of the vaccination programs, there is a general recognition that effective drugs are also needed, especially in view of the continuing threat of the sudden emergence of a major pandemic caused by a highly virulent mutant.

Amantadine and rimantadine

These closely related compounds (Figure 6.4) have both prophylactic and therapeutic activity against influenza A_2 infection, although they are ineffective against the less common type B virus. They prevent influenza A_2 infection as long as dosing is continued, and the duration of the disease appears to be shortened even if the drugs are started after the infection has begun. *In vitro,* the compounds inhibit replication of the influenza virus after the virus has entered the host cell but before the process of viral uncoating begins. The mature influenza virus particle is enveloped in a lipid membrane that contains three integral proteins: hemagglutinin, neuraminidase and a protein designated as M_2. This latter protein, which is a homo-oligomer, has been identified as the specific molecular target for amantadine. During a viral attack on susceptible cells, the virus particles, or virions, enter the cells by endocytosis and are then incorporated into the endosomal compartment. At this stage a tetramer of the M_2 protein functions as a highly selective proton channel across the virion membrane. A specific histidine residue in the M_2 protein (His-37) is essential for proton selectivity. A proton flux from the endosome through the channel into the virion interior ensures the reduction in internal pH that is required for the uncoating of the virion. The ion channel activity of M_2 has been confirmed with recombinant M_2 protein inserted into the membranes of *Xenopus* oocytes. The antiviral action of amantadine and rimantadine results from a specific interaction between the drugs and the M_2 protein which abrogates its ion channel function. Amantadine and rimantadine can be thought of as hydrophobically stabilized surrogates for H^+ ions that compete for proton binding at the lone electron pairs of His-37.

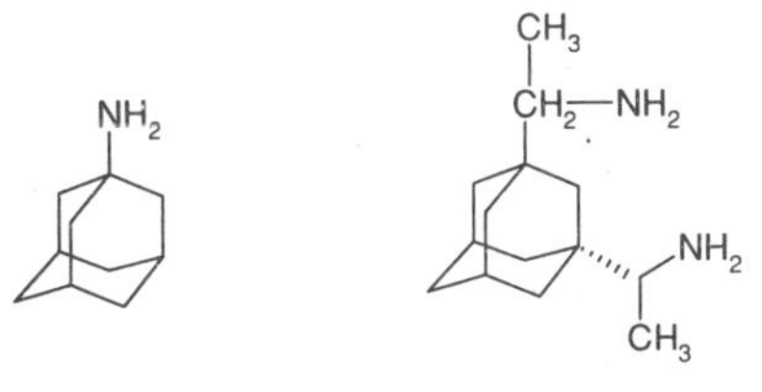

Amantadine　　Rimantadine

FIGURE 6.4 Two drugs active against influenza type A viruses with a unique action against the ion channel function of the viral protein M_2.

The usefulness of amantadine and rimantadine as antiinfluenza drugs is limited by their side effects, especially those affecting the central nervous system. Indeed, the primary medical use of amantadine is as an adjunct in the treatment of Parkinson's disease.

Inhibitors of viral neuraminidase in the treatment of influenza

Amantadine and rimantadine have never achieved widespread use for the treatment and prevention of influenza. An alternative approach has developed inhibitors of viral neuraminidase which have the advantage of being effective against both influenza type A and type B viruses. The two neuraminidase inhibitors currently available, zanamivir and oseltamivir (Figure 6.5), are most effective when taken as soon as possible

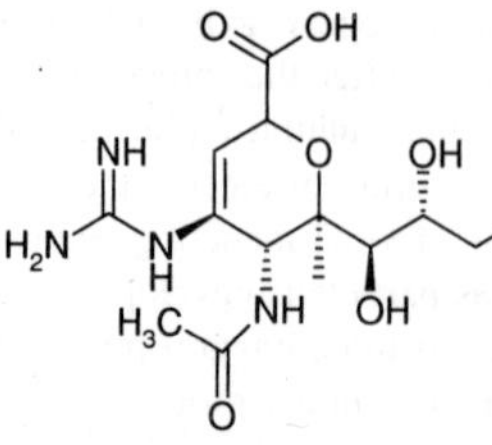

Zanamivir

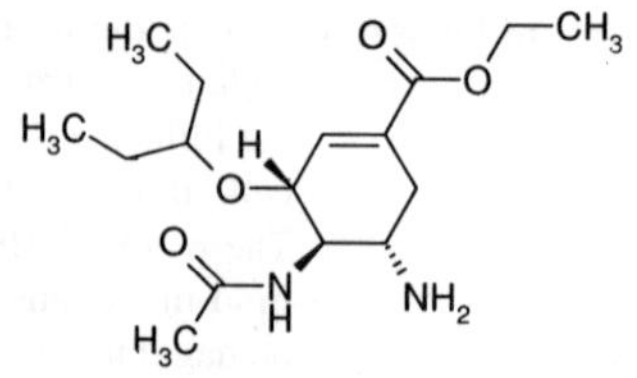

Oseltamivir

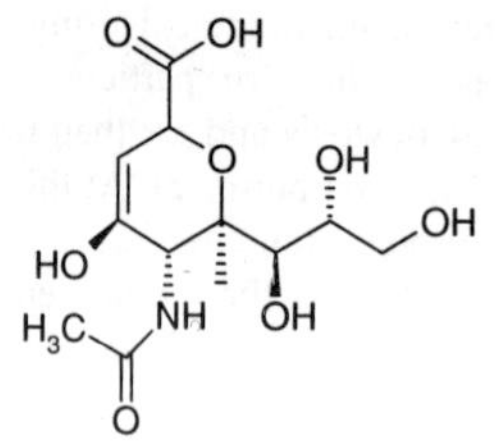

DANA: 2,3,-dehydro-2-deoxy-*N*-acetylneuraminic acid

FIGURE 6.5 Inhibitors of viral neuraminidase with activity in early-stage infections caused by influenza type A and type B viruses. DANA is a transition state intermediate in the catalytic action of the enzyme used in the design of inhibitors.

after the symptoms of 'flu first appear. The subsequent duration of the infection is significantly shortened and the risk of serious complications, such as bronchitis, is reduced.

The attachment of the influenza viruses to the surface of target cells and the subsequent release of progeny viruses is mediated successively by two glycoprotein spikes on the viral coat: hemagglutinin and neuraminidase. Hemagglutinin binds to the sialic acid moieties of surface receptors on the target cells and initiates viral adsorption and penetration. After replication inside the cells, progeny virions budding from the cell membrane bind via their hemagglutinin spikes to the cell surface receptors and to other virus particles. The viral neuraminidase now cleaves the terminal sialic residues from the receptors and releases the newly formed viruses from the cell membranes and from each other. Neuraminidase may also facilitate the migration of viruses through the mucin layer which covers the respiratory tract.

The elucidation of the three-dimensional structure of neuraminidase and its interaction with the cell surface receptors during the 1990s paved the way for the design of inhibitors of the enzyme. The structure of the first inhibitor, zanamivir, is analogous to that of a transition state analogue believed to be involved in the catalytic process, 2,3-dehydro-2-deoxy-*N*-acetylneuraminic acid (DANA, Figure 6.5). Zanamivir binds to amino acids (glutamic acid-119 and arginine-292) in the active center of the enzyme which are conserved in both type A and type B viruses. The cleavage of the terminal sialic acid residues of the cell receptors is inhibited, suppressing the release of progeny virions and their spread to other cells.

Zanamivir must be administered by inhalation into the respiratory tract. The structurally similar oseltamivir has the advantage of being effective if administered orally as an ethyl ester prodrug which is converted to the active inhibitor after absorption by liver esterases.

6.3.4 Interferon

Originally discovered in 1957 as a naturally occurring antiviral agent, interferon is now used as the generic name for a family of proteins involved in host defences against certain viral and parasitic protozoal infections. Interferons (IFNs) also affect the immune system, cell proliferation and differentiation and thus have a useful, although limited, antitumour activity. The therapeutic opportunities for the interferons have expanded considerably with the advent of technology to provide substantial quantities of recombinant proteins.

The nomenclature for IFNs, based on amino acid sequence data, defines four groups: IFN-α and IFN-ω (previously IFN-α-1 and IFN-α-2), IFN-β and IFN-γ. Only IFN-α is used therapeutically as an antiviral drug and it is valuable in the treatment of hepatitis B and C. IFN-α is also useful against hairy cell leukemia and AIDS-related Kaposi's sarcoma, actions that may depend upon its antiproliferative rather than its antiviral property. IFN-α itself constitutes a family of related proteins encoded by at least 14 functional genes.

The antiviral activity of IFN-α is mediated through a complex cascade of events, beginning with the binding of the protein to its specific receptor embedded in the cytoplasmic membrane of the cell. The IFN-α receptor consists of at least two subunits, IFN-AR1 and IFN-AR2. A current model of IFN-α receptor function proposes that the R1 and R2 subunits are associated, respectively, with two distinct protein tyrosine kinases, Tyk2 and JAK1 (a 'Janus' kinase). The binding of IFN-α to the receptor results in the activation of these kinases which then phosphorylate the gene transcription factors STAT1 and STAT2 on specific tyrosine residues. The activated STAT molecules form heterodimers that translocate to the nucleus in association with an additional factor, IFN regulatory factor. The resulting complex, termed IFN-stimulated gene factor-3, then interacts with the IFN-stimulated response element to modulate the transcription of more than 300 genes. To add to the complexity, there is also evidence for the involvement of additional transcriptional factors, STAT3 and STAT5.

Although many proteins may contribute to the antiviral action of IFN-α, which includes suppression of penetration, uncoating, transcription, translation and virus assembly, attention has largely focused on the roles of three key proteins induced by the signal transduction process:

1. 2′,5′-oligoadenylate synthetase, otherwise known as (2′-5′)(A_n) synthetase, which is active only in the presence of double-stranded RNA (dsRNA, an intermediate or byproduct of viral replication),
2. RNAase L, and
3. dsRNA-dependent protein serine kinase.

The (2′-5′)(A_n) synthetase exists in several isoenzymic forms and catalyzes the conversion of ATP to a series of AMP oligomers linked by 2′-5′ rather than the usual 3′-5′ phosphodiester bonds. The 2′-5′ A oligomers (up to 15 units in length) then activate the latent form of RNAase L. The active form of this endonuclease hydrolyzes both mRNAs and rRNAs at sequences containing UU and UA. The destruction of RNA molecules contributes to both the antiviral and antiproliferative actions of IFN-α.

As mentioned earlier, IFN-α is useful in the treatment of hepatitis C, usually in combination with ribavirin (Chapter 4). Recent evidence indicates that the hepatitis C virus interferes with the JAK-STAT signaling sequence induced by IFN-α, which may depress the effectiveness of the drug through reduced transcription of the IFN-α-stimulated genes.

6.4 Antiprotozoal agents

6.4.1 Antimalarial drugs

Quinolines

The statistics for malaria are appalling: some 2.5 billion people are at risk from the disease with 300–500 million new cases every year, of whom as many as 2 million will die. The first effective antimalarial drug was quinine (Figure 6.6), which is present in the bark of the cinchona tree of South America. Later chemically related but wholly synthetic drugs include chloroquine (Figure 4.14), mepacrine and mefloquine (Figure 6.6). Drugs that intefere with the folic acid metabolism of malarial parasites are described in Chapter 4.

Of the quinoline drugs, chloroquine has been a mainstay of both the prophylaxis and treatment of malaria for more than 60 years. Unfortunately, resistance to this drug is now widespread (Chapter 9). Early studies suggested that the antimalarial action of chloroquine might depend upon its ability to bind to DNA by intercalation (Chapter 4), leading to inhibition of DNA replication and transcription. However, although it is still possible that the intercalative property of chloroquine contributes to its antimalarial

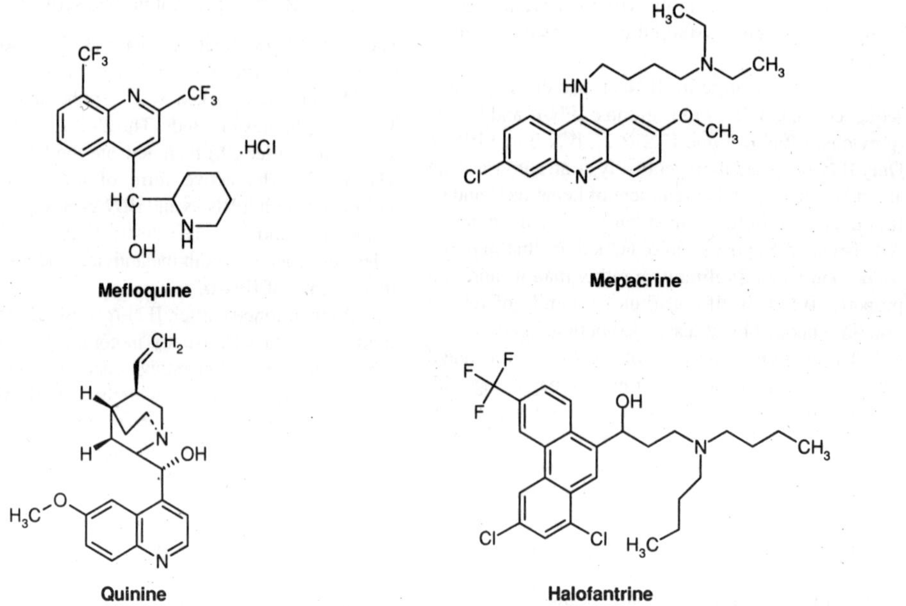

FIGURE 6.6 Four well-established antimalarial drugs.

action, the principal effect of the drug is to disrupt the ability of the parasite to cope with heme released during the metabolism of hemoglobin.

At a certain stage in the complex life cycle of *Plasmodium* parasites, the protozoans invade the red cells of the host. The trophozoites, as the parasitic cells are called at this stage, digest more than 80% of the hemoglobin of the infected red cells in lysosomal vacuoles to obtain amino acids for their own development. The digestive process releases the cytolytic porphyrin heme within the parasitic cell. The trophozoites protect themselves against the toxic effect of heme in two ways. One way is by facilitating polymerization of heme into an inert, insoluble crystalline substance called hemozoin, or β-hematin. In this molecule the iron atom of one molecule of heme is coordinated to the propionate carboxyl residue of the next heme. The resulting dimers form chains linked by hydrogen bonds. There is still debate as to whether this polymerization and crystallization process is mediated by an enzyme or protein of the trophozoites or whether it is a spontaneous reaction dependent solely on the chemical nature of heme. In the second method, a significant proportion of the heme released by the digestion of hemoglobin escapes incorporation into hemozoin. The residual heme is oxidatively degraded by hydrogen peroxide formed as a by-product of the oxidation of Fe^{2+} to Fe^{3+} immediately following its release from hemoglobin. Another fraction of heme is degraded by interaction with glutathione catalyzed by glutathione reductase.

The positive charge on the chloroquine molecule is probably responsible for its accumulation within the acidic environment of the food vacuoles of the parasite to concentrations as high as 100 μM. There are several histidine-rich proteins (HRPs) in the digestive vacuoles of *Plasmodium falciparum* that promote the polymerization of heme. One of these proteins, HRP-II, has 51 His-His-Ala repeats that may provide the binding sites for the 17 molecules of heme that bind to 1 molecule of HRP-II. Chloroquine inhibits the HRP-II-mediated formation of hemozoin. However, against this must be set the finding by other investigators that chloroquine also inhibits hemozoin formation in the absence of any added proteins. Whether hemozoin synthesis is protein mediated or is a spontaneous phenomenon within the trophozoites, it seems likely that the inhibitory effect of chloroquine on the process depends on an interaction between the drug and the heme molecule or minimal heme oligomers.

Chloroquine also competitively inhibits the degradative processes that eliminate heme not captured by hemozoin formation. The drug may form a complex with heme that blocks peroxidation by hydrogen peroxide and its reaction with glutathione. Toxic chloroquine-heme complexes accumulate and contribute to the death of the parasitic cells.

Despite the chemical similarity of the quinoline drugs shown in Figure 6.6, there are differences in their modes of antimalarial action. While chloroquine, mepacrine and mefloquine all inhibit hemozoin formation, the inhibitory actions of mefloquine and quinine on the ability of the protozoan to accumulate hemoglobin in its digestive vacuoles may also contribute to the antimalarial action of these drugs. After many years of effort, the final details of the antimalarial actions of the quinoline drugs still remain to be defined. As we have seen, the drugs inhibit several aspects of trophozoite function which would contribute to a synergistic attack on cell survival.

Although halofantrine (Figure 6.6) is not a member of the quinoline group, its prinicipal mode of antimalarial action is probably broadly similar to that of the quinolines, i.e. it blocks the elimination of the cytolytic heme. There is *in vitro* evidence that halofantrine inhibits the glutathione-mediated degradation of heme. Halofantrine has potentially dangerous side effects on cardiac function which may be associated with drug-mediated inactivation of the essential potassium channels in the cytoplasmic membrane of myocardial cells. It is conceivable that interference with ion channel function, such as the proton pump of the protozoan food vacuoles, could also contribute to the antimalarial action of halofantrine.

Artemisinin

An infusion of the leaves of the plant *Artemisia annua* is an ancient Chinese herbal remedy for fevers, including malaria. The active principle, artemisinin (Figure 6.7), is highly active against the malarial parasite. Artemisinin and its more convenient water-soluble derivative, artesunate (Figure 6.7), are proving to be useful new weapons in the fight against malaria, particularly against the chloroquine-resistant and cerebral forms of the disease. These drugs rapidly kill all the asexual stages of the most dangerous malarial parasite, *Plasmodium falciparum*. Artemisinins are usually given in combination with inhibitors of folic acid metabolism or with mefloquine.

Artemisinin and its close structural analogues diffuse readily into red cells infected by the parasite. Infected red cells are found to contain up to 100 times more artemisinin than uninfected cells. The endoperoxide bridge of the drug molecules then reacts with the ferrous iron atom of heme released from digested hemoglobin. The chemistry of the interaction of artemisinin with heme and the susbequent generation of reactive carbon-centered free radicals is complex, and further details can be found in a reference in 'Further reading.' However, the essential role of the endoperoxide bridge is evidenced by the fact that derivatives lacking this feature are inactive. Molecular modeling studies suggest that the endoperoxide bridge can achieve close proximity to the heme iron. The interaction with the Fe^{3+} atom catalyzes the breakdown of artemisinin into a complex cascade of unstable intermediates, the details of which are still debated. The balance of evidence suggests that a carbon-centered

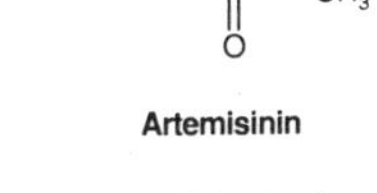

Artemisinin

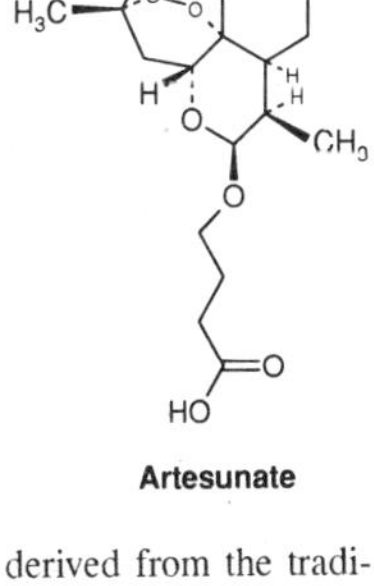

Artesunate

FIGURE 6.7 Antimalarial agents derived from the traditional Chinese herbal medicine *Artemisia annua*, usually given in combination with other antimalarial drugs.

radical is generated which alklylates essential proteins and leads to the death of the parasite. However, it is also possible that active oxygen species are generated during the breakdown of artemisinin, which would cause peroxidation of lipids in the vacuolar membrane of the parasite and subsequent dissolution of the vacuole.

Recently a more specific explanation has been advanced for the antimalarial action of the artemisinins. The carbon-centered free radical generated by the interaction of artemisinin with heme is revealed as a potent inhibitor of the Ca^{2+}-dependent ATPase (SERCA) of the sarcoplasmic and endoplasmic reticulum of the protozoan cells. This enzyme, referred to as PfATP6, is not inhibited by quinine or chloroquine. A direct link between the enzyme inhibition of PfATP6 and cell death has not been established, although there is a correlation between the inhibitory potency of a range of artemisinin derivatives and their ability to kill the parasite. The alkylation and enzyme inhibition of PfATP6 may well synergize with the effects of the free radical on other, as yet undefined, proteins to achieve the lethal action of artemisinins. The malarial parasite develops resistance to artemisinin-type drugs with considerable difficulty, which further suggests that the antimalarial action of these drugs is multifactorial.

Atovaquone

Although hydroxyanthraquinones have been known for 60 years or more to inhibit the respiration of malarial parasites, the introduction of the hydroxynapthaquinone derivative atovaquone (Figure 6.8) for the treatment of malaria is relatively recent. In combination with an inhibitor of dihydrofolate reductase, such as proguanil (Figure 4.3), atovaquone is highly effective in the prevention of malaria caused by *Plasmodium falciparum*. The combination is largely free of the side effects that can make compliance with other antimalarial drug regimes difficult. Atovaquone is also used to treat another protozoal infection caused by *Toxoplasma gondii* and the dangerous respiratory fungal pathogen *Pneumocystis carinii*.

The primary inhibitory action of atovaquone is against the respiratory chain in mitochondria, leading to an interruption in the supply of ATP. The target site is the ubiquinone-cytochrome *b* and *c* reductase region of the respiratory chain, i.e. the cytochrome bc_1 complex. The complex catalyzes electron transfer from ubiquinone to cytochrome *c* and at the same time translocates protons across the mitochondrial membrane. The complex contains prosthetic groups, cytochrome *b*, cytochrome c_1 and a nonheme iron-sulfur protein. A partial structural similarity between atovaquone and ubiquinone suggests that there is competition between the two molecules at the cytochrome bc_1 complex. Because of the technical problems associated with culturing malarial parasites *in vitro*, it is difficult to isolate mitochondria from the parasites in sufficient amounts to investigate the interaction between atovaquone and its target site. Fortunately, there is a high degree of amino acid sequence similarity between the cytochrome *b* of *Plasmodium falciparum* and that of the yeast *Saccharomyces cerevisiae*. The cytochrome bc_1 complex is readily purified from the yeast and the details of its crystal structure are known.

Structural modeling with the yeast preparation revealed that atovaquone potentially competes for the occupation of the ubiquinol oxidation pocket, of the bc_1 complex. The soluble domain of the iron-sulfur protein is proximal to the oxidation pocket, and recent molecular modeling studies based on the crystal structure of the bc_1 complex, together with biochemical spectroscopic experiments, point to hydrogen bonding between the hydroxyl group of the drug and histidine-181 of the iron-sulfur protein. The quinone carbonyl group of the drug is seen to hydrogen bond with glutamate-272 of the cytochrome *b* component when a bound water molecule is introduced into the model to form a bridge between the quinone carbonyl group and glutamate-272.

The model of the interaction of the atovaquone-cytochrome bc_1 complex provides some insight into the potential for the selective inhibition of mitochondrial function in the microbial parasites. The ef loop of cytochrome *b* contains amino acid residues in close contact with the atovaquone-binding pocket. Position 275 in the loop is occupied by leucine in *Saccharomyces cerevisiae* and in *Pneumocycstis carinii*, but is replaced by phenylalanine in the human protein. Modeling of this replacement indicates that phenylalanine causes significant steric hindrance to the binding of atovaquone, which could account for the

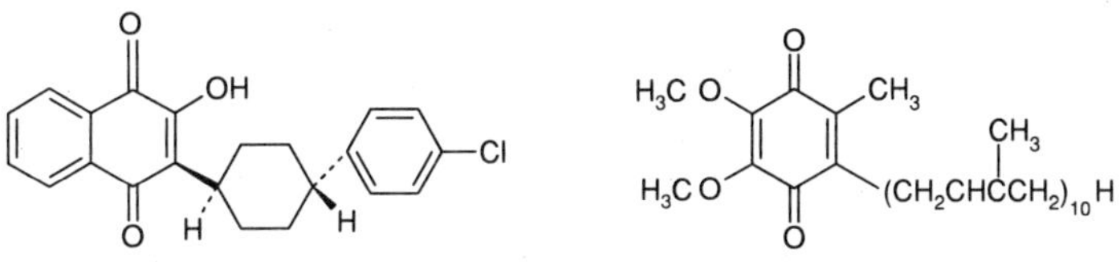

Atovaquone Ubiquinone

FIGURE 6.8 Atovaquone, an inhibitor of protozoan mitochondrial function recently introduced into the prophylaxis and therapy of malaria in combination with inhibitors of folic acid metabolism. Ubiquinone is shown here to illustrate its partial structural resemblance to atovaquone.

markedly lower potency of the drug against mammalian mitochondria.

6.4.2 Antitrypanosomal drugs

There are two major forms of trypanosomiasis: African sleeping sickness, caused by two species of *Trypanosoma brucei,* and Chagas' disease or South American trypanosomiasis, caused by *Trypanosoma cruzi*. African sleeping sickness has become resurgent in sub-Saharan Africa, causing an estimated 100,000 deaths each year. Chagas' disease affects some 18 million people in South America, with about 25% of the population at risk of acquiring the disease. Unfortunately there are few drugs for the treatment of either form of trypanosomiasis and most of them were developed decades ago. Research into the chemotherapy of trypanosomiasis remains relatively neglected. The action of another antitrypanosomal drug, ethidium, used in veterinary medicine, is discussed in Chapter 4.

Suramin

First made available over 70 years ago, suramin (Figure 1.2) is a large, polysulfonated molecule with six negative charges at physiological pH. It is often the drug of choice for the treatment of the early stages of African sleeping sickness. In view of its molecular size and highly charged nature, it is surprising that suramin is able to gain access to the cytoplasm of the parasitic cells. In fact, the compound binds avidly to serum proteins and it is believed that both free proteins and those complexed with suramin enter trypanosomes by endocytosis. The selective toxicity of suramin for trypanosomes may in part be due to a more avid uptake of serum proteins by trypanosomes compared with the cells of the infected patient. The concentration of suramin in trypanosomes can reach 100 μM.

Suramin binds to and inhibits a broad range of enzymes derived from *Trypanosoma brucei,* including dihydrofolate reductase, thymidine kinase and all the enzymes of the glycolytic pathway. Although the IC_{50} values (the concentration of inhibitor needed to inhibit an enzyme by 50%) are in the high range of 10–100 μM, they are much lower than those for the corresponding enzymes from mammalian sources. The glycolytic enzymes of the trypanosome, which generate all of its ATP, are confined to membrane-bounded organelles called glycosomes. Although the physical characteristics of suramin make it unlikely to diffuse into glycosomes, the drug probably binds to the newly synthesized proteins during their cytoplasmic phase prior to entry into the glycosomes. The normal turnover of uncomplexed enzymes in the glycosomes would be expected to lead to their gradual replacement by suramin-bound enzymes entering from the cytoplasm. This model is consistent with the observed progressive slowing down of energy metabolism in cells treated with suramin. The antitrypanosomal action of suramin is unlikely to depend on the inhibition of a single enzyme, a conclusion that is supported by the fact that resistance to suramin has not been a serious problem despite many decades of use. A multifaceted mode of action probably hinders the ability of trypanosomes to develop resistance to the drug.

Pentamidine

This diamidine compound (Figure 6.9), like suramin, is useful in treating the early stages of African trypanosomiasis. By exploiting a protozoal aminopurine transport system, pentamidine is concentrated within the trypanosomes. Despite being relatively nonspecific in its interaction with macromolecules, pentamidine probably exerts its primary antitrypanosomal action by attacking and selectively cleaving the DNA of the kinetoplast.

Melarsoprol

Otherwise known as melarsen oxide, melarsoprol (Figure 6.9) is a trivalent organic arsenical drug that has been used to treat sleeping sickness since 1949. Unlike suramin and pentamidine, it is useful against late-stage disease, although treatment is fraught with the risk of serious side effects, especially a potentially lethal encephalopathy. Melarsoprol also enters the trypanosome via an aminopurine transporter.

Although African trypanosomes incubated with melarsoprol die within minutes, the mechanism of action is not clear and may have more than one aspect to it. Several glycolytic enzymes are inhibited by melarsoprol, including phosphofructokinase ($K_i < 1$ μM), fructose-2,6-diphosphatase (i.e. $K_i = 2$ μM) and, to a lesser extent, pyruvate kinase ($K_i > 100$ μM). Melarsoprol also disrupts the function of a unique trypanosomal biochemical, trypanothione [N^1,N^8-*bis*(glutathionyl)-spermidine], with which the drug forms a stable adduct. Trypanothione is a major cofactor in the control of the redox balance between thiols and disulfides in trypanosomes. The adduct with melarsoprol inhibits trypanothione reductase ($K_i = 17.2$ μM), which is a key enzyme regulating the redox state of trypanothione itself. However, the relatively high K_i casts some doubt on the relevance of the inhibition of trypanothione reductase to the antitrypanosomal action of melarsoprol. Furthermore, incubation of *Trypanosoma brucei* cells with the drug leads to only a minor conversion of reduced trypanothione to its adduct with melarsoprol.

Melarsoprol is typical of organic arsenical agents in its ability to form adducts with thiols and may therefore inhibit many enzymes with essential thiol groups or that require thiol-containing cofactors. Thus although the major effect of melarsoprol is probably to suppress glycolysis in African trypanosomes, the inhibition of other enzymes may well contribute to the overall antitrypanosomal action. The toxicity of melarsoprol to the patient is also almost certainly due to the avidity of the drug for thiols.

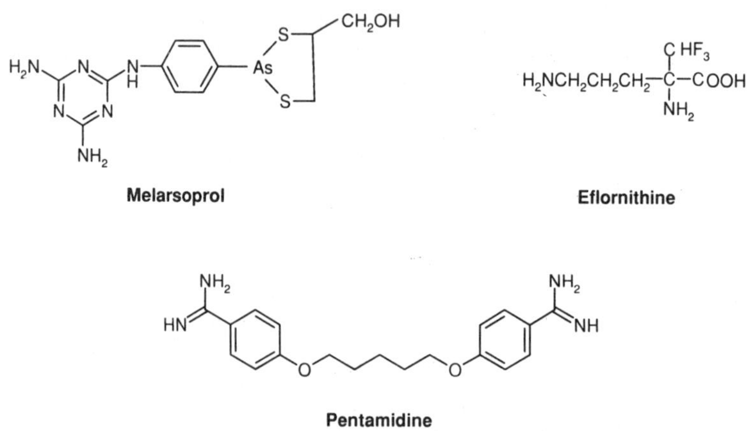

FIGURE 6.9 Drugs used to treat African trypanosomiasis.

Eflornithine

DL-α-difluoromethylornithine (DFMO) or eflornithine (Figure 6.9) was originally designed as a 'suicide' inhibitor of ornithine decarboxylase (ODC) for use as an antitumour drug. ODC is involved in the biosynthesis of the polyamines putrescine, spermidine and spermine, which are essential for cell division in eukaryotic cells. Depletion of cellular polyamines caused by the inhibition of ODC results in the suppression of mitosis and cell proliferation. In addition to its potential as an anticancer agent, eflornithine has also proved to be an effective drug against African trypanosomiasis. However, adverse side effects, which are similar to those encountered with other cytotoxic drugs, are common during antitrypanosomal therapy with eflornithine. Fortunately these effects are reversible after treatment ceases. When given in high doses for 14 days, eflornithine is effective against both early- and late-stage disease.

The basis of the trypanosome-selective action of eflornithine hinges on marked differences in the turnover rates of ODC in mammals and the trypanosome. Eflornithine is an irreversible inhibitor of mouse ODC, forming a covalent adduct with cysteine-360. Since human ODC shares a 99% sequence identity with the mouse enzyme, it can be reasonably assumed that the human enzyme is also inhibited by eflornithine in the same way. The K_i against mouse ODC is 39 μM, compared with 220 μM against the corresponding enzyme from *Trypanosoma brucei*. This might lead one to expect that the drug would be more effective against mouse and mammalian cells in general than against trypanosomes. However, while the trypanosomal enzyme is highly stable with minimal intracellular turnover, mammalian ODC has a half-life of only 20 min, placing it among the most rapidly metabolized of eukaryotic proteins. This, combined with the rapid elimination of eflornithine from the body, results in a single dose of the drug exerting only transient inhibition of the constantly renewed ODC of the host. By contrast, there is sustained inhibition of the trypanosomal ODC, resulting in depletion of putrescine and spermidine as well as of trypanothione (recall that the latter is a conjugate of glutathione with spermidine). The drug-treated trypanosomes cease dividing and become incapable of changing their variant surface glycoprotein (VSG). Normally, continual changes in VSG provide the basis of the remarkable ability of trypanosomes to evade immunological detection and destruction by the host. When changes in the VSG are prevented, the parasite becomes vulnerable to immunological attack, which assists in the resolution of the infection.

The antitrypanosomal drugs described here are effective only against the African forms of trypanosomiasis. There are even fewer drugs effective against Chagas' disease. The preferred approach is prevention of the disease by controlling the insect vector for *Trypanosoma cruzi*, the reduviid bug, which infests poorly constructed housing. Nevertheless, several drugs are being tested against Chagas' disease, the best of which is the nitroheterocyclic compound benznidazole (Figure 6.1). The action of this drug is similar to that of the compounds discussed in Section 6.1, i.e. it is reductively activated in the protozoan cell to generate a free radical which attacks DNA and other macromolecules in the cells. However, full details of the action of benznidazole in *Trypanosoma cruzi* have not been defined. Treatment of Chagas' disease with benznidazole and other nitroheterocyclic drugs is far from satisfactory, owing to the risk of DNA damage to the patient during the sustained doses needed to eliminate the parasite.

Further reading

Chaudhuri, A. R. and Ludeña, R. F. (1996). Griseofulvin: a novel interaction with brain tubulin. *Biochem. Pharmacol.* **51**, 903.

Cohen, J. L. (1996). Protease inhibitors: a tale of two companies. *Science* **272**, 1882.

Dachs, G. V., Abatt, V. R. and Woods, D. R. (1995). Mode of action of metronidazole and a *Bacteroides fragilis metA* resistance gene in *Escherichia coli*. *J. Antimicrob. Chemother.* **35**, 483.

Duong, F. H. (2004). Hepatitis C virus inhibits interferon signaling through up-regulation of protein phosphatase 2A. *Gastroenterology* **126**, 263.

Egan, T. J., Ross, D. C. and Adams, P. A. (1994). Quinoline antimalarial drugs inhibit spontaneous formation of β-hematin (malaria pigment). *FEBS Lett.* **352**, 54.

Ekstein-Ludwig, U. *et al.* (2003). Artemisinins target SERCA of *Plasmodium falciparum*. *Nature* **424**, 957.

Jefford, C. W. (2001). Why artemisinin and certain synthetic peroxides are potent antimalarials: implications for the mode of action. *Curr. Med. Chem.* **8**, 1803.

Kerr, I. M. *et al.* (2003). Of JAKs, STATs, blind watchmaker, jeeps and trains. *FEBS Lett.* **546**, 1.

Kessl, J. L. *et al.* (2003). Molecular basis for atovaquone binding to the cytochrome bc_1 complex. *J. Biol. Chem.* **278**, 31312.

Lear, J. D. (2003). Proton conduction through the M2 protein of the *influenza A* virus: a quantitative, mechanistic analysis of experimental data. *FEBS Lett.* **552**, 17.

Matthews, T. *et al.* (2004). Enfuvirtide: the first therapy to inhibit the entry of HIV-1 into host CD4 lymphocytes. *Nat. Rev. Drug Discov.* **3**, 215.

McKimm-Breschkin, J. L. (2002). Neuraminidase inhibitors for the treatment and prevention of influenza. *Expert. Opin. Pharmacother.* **3**, 103.

Menéndez-Arias, L. (2002). Targeting HIV: antiretroviral therapy and development of drug resistance. *Trends Pharmacol. Sci.* **23**, 381.

Raether, W. and Hänel, H. (2003). Nitroheterocyclic drugs with broad spectrum activity. *Parasitol Res.* **90**, S19.

Sanchez, C. P. and Lanzer, M. (2000). Changing ideas on chloroquine in *Plasmodium falciparum*. *Curr. Opin. Infect. Dis.* **13**, 653.

Urbina J. A. (2002). Chemotherapy of Chagas' disease. *Curr. Pharm. Des.* **8**, 287.

Chapter seven

Attack and defense: drug transport across cell walls and membranes

To inhibit microbial growth, drugs must reach and maintain inhibitory concentrations at the target sites which usually reside in the cytoplasm or are embedded in the cytoplasmic membrane. Antimicrobial agents traverse the permeability barriers provided by cell walls and membranes that separate the target sites from the external environment by passive, or in some cases, facilitated diffusion. In addition to structural barriers hindering drug access to their targets, many wild-type micro-organisms actively extrude inhibitory compounds from the cytoplasm back into the external environment via a battery of drug efflux pumps. The structural permeability barriers and drug efflux systems combine to modulate the level of intrinsic resistance to drugs in wild-type organisms. An isolated target site prepared from different species of micro-organism may have sensitivities comparable to a specific drug *in vitro*, whereas the intact cells may show very different responses to the same drug. Higher levels of resistance acquired during sustained exposure to drugs are caused by a range of genetically based mechanisms discussed in later chapters, including reduced drug access that is due to increased cell wall thickness, loss of porins from the outer membranes of Gram-negative bacteria and increased levels of expression of drug efflux pumps. The aggregation of bacterial cells in some situations to form biofilms may lead to reduced susceptibility to drug action, and this has been attributed to hindered drug access through the biofilm. For example in patients with cystic fibrosis, *Pseudomonas aeruginosa* forms biofilms in lung which are notoriously difficult to treat. However, the present consensus is that the altered dynamics of bacterial cell growth in biofilms probably accounts for the lower drug sensitivity and that biofilms do not in general pose a significant diffusional barrier to drug access.

7.1 Cellular permeability barriers to drug penetration

7.1.1 The cytoplasmic membrane

The permeability barrier provided by the cytoplasmic membrane of micro-organisms depends on the characteristic lipid bilayer that is common to all biological membranes. Drugs cross this barrier either by passive diffusion or by facilitated diffusion involving a biological carrier system.

Passive diffusion

The rates of passive diffusion of uncharged organic molecules across lipid membranes are governed by Fick's law of diffusion and correlate reasonably well with their lipid/water partition coefficients. Fick's law is expressed by the equation:

$$V = P \times A(S_e - S_i)$$

where V is the rate of diffusion (in nmol $mg^{-1}s^{-1}$), A is the surface area of the membrane (cm^2mg^{-1}), S_e and S_i are, respectively, the external and internal concentrations of free permeant, and P is the permeability coefficient (cm s^{-1}). The internal concentration of free drug can be substantially affected by binding to intracellular macromolecular targets, metabolism to other chemical species or by changes in ionization that are due to differences between internal and external pH values. Lowering the internal free concentration of a compound by any of these factors enhances the rate of inward passive diffusion by steepening the concentration gradient of unbound drug. Diffusion across membranes is bi-directional, and the rates at which a compound diffuses into and out of a cell determine the time at which a steady-state intracellular concentration of the compound is achieved. In reality, however, a steady-state intracellular drug concentration in an infecting micro-organism may be achieved only transiently, if at all, because of rapidly changing conditions both inside the cell and in its external environment.

The greater the lipid solubility of a compound, expressed as the partition coefficient, the more readily it enters and diffuses across the lipid bilayer of the membrane. However, when lipid solubility is so high that a compound is essentially insoluble in water, it may be unable to diffuse out of the lipid interior of the membrane into the aqueous environment of the cytoplasm. This adversely affects the biological activity of compounds with sites of action within the cytoplasmic compartment but could enhance activity if the target site is membrane-bound. The relationship between biological activity and lipophilicity is expressed in the 'Hansch equation' (so named after the scientist who formulated it):

$$\log (1/C) = -k(\log P)^2 + k' \log P + \rho\sigma + k'',$$

where C is the molar concentration of the drug for a standard biological response, in the case of antimicrobial drugs usually the minimal inhibitory concentration (MIC) or alternatively, the concentration needed for 50% inhibition of growth or cell survival (IC_{50}); P is the partition coefficient; ρ and σ are physicochemical constants (Hammett constants) defining certain electronic features of the molecule; and k, k' and k'' are empirically determined constants. The equation indicates that within a chemically related series of biologically active molecules having similar values of ρ and σ, an optimal partition coefficient is associated with maximum biological activity. However, this relationship holds only for those agents that cross membranes by passive diffusion and it may break down when biologically facilitated transport is involved or when permeation occurs through water-filled pores, as in the case of the porin channels in the outer membrane of Gram-negative bacteria. Futhermore, the lipophilicity of a drug may be the critical factor determining its affinity for the target site because of the dominant contribution of hydrophobic forces to the drug–target interaction.

The Hansch equation has been applied to sets of synthetic antibacterial compounds that penetrate the bacterial envelope by passive diffusion. The results show that the compounds most active against Gram-negative bacteria are generally less lipophilic (or alternatively, more hydrophilic) than compounds with primary activity against Gram-positive organisms. The cytoplasmic membranes of the two classes of bacteria are sufficiently similar to make it unlikely that they could account for the differences in the partition coefficients of optimally active compounds. The explanation lies largely in the unique properties of the porin channels in the Gram-negative outer membrane, which facilitate the influx of hydrophilic compounds. We shall return to this important topic later in the chapter.

The rates of passive diffusion of water-soluble molecules across lipid membranes are usually very low, although uncharged polar compounds with molecular masses of less than 100 Da diffuse more readily. Nevertheless, as we shall see, certain hydrophilic antimicrobial agents of much higher molecular mass readily enter the cytoplasm. Ionized compounds diffuse across cytoplasmic membranes with difficulty, unless the molecules contain compensatory lipophilic regions, because the strongly bound hydration shells of ionized groups in aqueous solution hinder diffusion across the lipid bilayer. The effect of ionization on the activity of an antibacterial agent is well illustrated by erythromycin. The pK_a of the basic dimethylamino group of this antibiotic is 8.8 and the concentration required for antibacterial activity decreases markedly as the pH of the bacterial medium is increased from neutrality towards 8.8. Presumably only un-ionized erythromycin molecules, which represent an increasing proportion of the total erythromycin as the pK_a of the drug is approached, diffuse into the bacteria.

Facilitated diffusion

A remarkable feature of cytoplasmic membranes is their ability to transfer certain ions, nutrients, waste products and toxins at much higher rates than are possible by passive diffusion. This process is known as facilitated transfer or facilitated diffusion. Characteristically, the rate of transfer of the permeant is proportional to its concentration over a limited range, beyond which a limiting rate is approached. This is due to the involvement of carrier proteins within the membrane that transiently bind the permeants and 'shuttle' them across the membrane. The rate of transfer increases with increasing permeant concentration until all of the carrier sites are saturated. This is in contrast to passive diffusion, where the transfer rate is proportional to permeant concentration over a much wider range. The kinetics of facilitated diffusion are directly comparable with the Michaelis-Menten kinetics of enzymes and their substrates. The following equation enables the transfer rate across the cytoplasmic membrane, v, to be calculated for a given concentration, C, of substance S:

$$v = V_{max}/(1 + K_m/C),$$

where C is the concentration of S either inside or outside the cell, depending on the direction the solute is being transferred; V_{max} is the maximum rate of diffusion when all the carrier sites are occupied by S and K_m is the concentration of S at which half the maximum number of carrier sites are occupied. K_m is therefore correlated with the affinity of the carrier molecule for S.

Facilitated transfer by itself results in the equilibration of the permeant across the membrane. However, when the transfer system is linked to an input of 'energy', usually the hydrolysis of ATP or the proton motive force across the cytoplasmic membrane, the permeant is transferred across the membrane against its concentration gradient. This is known as active transport. Some facilitated transfer systems are highly specific and only close structural analogues of the natural permeant compete effectively for the transport sites. In contrast, most drug efflux pumps exhibit remarkably broad specificity in the range of compounds they remove from microbial cells (see later discussion).

7.1.2 The cell walls of bacteria and fungi

The function of the peptidoglycan cell wall of bacteria and of the multilayered glycoprotein-polysaccharide fungal cell wall is to give shape and tensile strength to microbial cells. These structures are mainly open networks of macromolecules and generally do not offer significant permeability barriers to compounds of molecular mass less than 50 kDa. Even the thick peptidoglycan walls of Gram-positive bacteria are permeable to antimicrobial peptides such as nisin and defensin, which have a molecular mass of 3 kDa or more. However, there are two important exceptions to the general rule of high permeability of bacterial cell walls:

1. As discussed in Chapter 2, the cell wall of a Gram-positive mycobacteria is characterized by a high lipid content that is due to the presence of long-chain mycolic acids on the outer surface of the wall which are covalently linked to the underlying arabinogalactan. The extended hydrocarbon chains of the mycolic acid molecules are tightly packed in a parallel array perpendicular to the cell surface. The outermost surface of the mycobacterial cell also contains a range of other complex lipids and waxes. This assembly has very low fluidity and resembles the outer membrane of Gram-negative bacteria. The mycobacterial cell wall is therefore a formidable permeability barrier to antibacterial drugs and accounts

for the well-known resistance of *Mycobacterium tuberculosis* to many antibiotics. The important antituberculosis antibiotic rifampicin (Chapter 4) is relatively hydrophobic and may gain access to its target in the cytoplasm by diffusion though the lipid barrier. Hydrophilic antituberculosis drugs such as isoniazid and ethambutol that interfere with the biosynthesis of the mycobacterial cell wall could initially diffuse through porin channels which are thought to penetrate the outer membranelike structure. Subsequent disruption of cell wall biosynthesis may then further facilitate influx of a drug.

2. Clinically significant acquired resistance of *Staphylococcus aureus* to vancomycin is associated with a greatly thickened cell wall, which it is suggested may trap the antibiotic within the extended peptidoglycan meshwork, hindering its access to the target sites (see Chapter 9).

Because of their strongly polar, predominantly negatively charged nature, the teichoic acids of Gram-positive cell walls could, in principle, influence the penetration of ionized molecules. The interaction of water-soluble, positively charged compounds, such as the aminoglycosides, with teichoic acid might generate locally high drug concentrations within the envelope, enabling the drugs to challenge the permeability barrier of the cytoplasmic membrane more effectively. In contrast, the entry of anionic molecules could be retarded by teichoic acid, although the exquisite sensitivity of many wild-type Gram-positive bacteria to penicillins, which are organic anions, shows that the repulsive effect of teichoic acid is not significant. It is unlikely that the teichoic acids of Gram-positive bacterial cell walls have any significant effect on the steady-state intracellular concentrations of antibiotics.

The Gram-negative outer membrane

In contrast to the thin peptidoglycan cell wall of Gram-negative bacteria, the outer membrane of these organisms is a significant contributor to the greater intrinsic resistance of Gram-negative bacteria to many antibiotics (Table 7.1). An indication that the outer membrane hinders drug penetration came from studies with

TABLE 7.1 Differential sensitivity to typical antibacterial drugs

Drugs active against Gram-positives and Gram-negatives	*Drugs less active against Gram-negatives*
Tetracyclines	Benzylpenicillin
Streptomycin and aminoglycosides	Methicillin
Sulfonamides	Macrolides
D-Cycloserine	Lincomycin
Chloramphenicol	Rifamycins
Fosfomycin	Fusidic acid
Many synthetic antiseptics	Vancomycin
Nitrofurans	Novobiocin
Ampicillin and carbenicillin	Bacitracin
Thienamycin	
Fluoroquinolones	

Gram-negative cells with defective envelopes. L-Phase (or L-forms) of *Proteus mirabilis* were found to be 100 to 1000 times more sensitive than intact cells to erythromycin and several other macrolides. There was a smaller increase in sensitivity to other antibiotics, including streptomycin, chloramphenicol and the tetracyclines. Both the outer membrane and the peptidoglycan are also defective in L-forms, so that the relative contributions of the various outer layers of intact bacteria to the barrier function were not certain in this early work. However, later studies clearly defined the outer membrane as a significant permeability barrier to some molecules.

Two major features of the Gram-negative outer membrane distinguish it from the cytoplasmic membrane:

1. Negatively charged lipopolysaccharide (LPS) in the outer leaflet of the bilayer replaces the glycerophospholipid of most other biological membranes. The negative charge of the LPS is partly neutralized by divalent cations, mainly Mg^{2+} and Ca^{2+}, which are readily removed by chelating agents such as ethylenediaminetetraacetic acid (EDTA).
2. Molecular diffusion across the complex lipid-lipopolysaccharide bilayer of the outer membrane is slow. In order to permit rapid influx of essential nutrients and ions, the

outer membrane is studded with water-filled pores formed by porin proteins. The porin channels, which are unique to the outer membrane, are instrumental in permitting the initial influx of certain hydrophilic antibacterial compounds across the outer envelope. Porin proteins consist of trimers of β-barrels arranged as antiparallel β-strands, commonly 16 in number, threading through the outer membrane. At the inner face of the membrane, the strands are joined by short β-turns and at the outer face by longer loops of amino acids. Three types of porin channel have been identified:

a. general channels with low permeant selectivity,
b. permeant-selective channels with internal specific binding sites, and
c. permeant-selective 'gated' channels that only open upon the binding of the specific permeant.

All three types of porin channel restrict transit to compounds with molecular masses of less than approximately 600 Da. The general OmpF porin of *Escherichia coli* is the most thoroughly studied porin and was the first membrane protein to be successfully crystallized for X-ray analysis. It is closely homologous to two other porins, PhoE, which allows the influx of phosphate ions, and OmpC. OmpF has pore dimensions (at its narrowest) of 11 × 7Å which allows the passage of major nutrients and hydrophilic antibiotics with molecular masses of < 600 Da. Although water-soluble antibiotics usually pass through the general channels, the selective porins are also used. The β-lactam antibiotic imipenem, for example, diffuses through the basic amino acid-specific channel OprD in *Pseudomonas aeruginosa.* The movement through the general porin channels of β-lactam antibiotics close to the molecular mass limit is retarded by repulsive interactions between the drugs and the predominantly negatively charged amino acids lining the channels. These interactions may hinder diffusion by as much as 100-fold. The diffusion of lipophilic molecules through the porin channels is much more difficult because the charged amino acid residues lining the narrowest regions of the channels orient their associated water molecules in a direction that hinders the passage of lipophilic permeants. The importance of porin channels to the influx of hydrophilic antibacterial agents is clearly demonstrated by the reduced susceptibility of porin-deficient mutants to antibiotics, including some valuable semisynthetic β-lactams.

In several species of Gram-negative bacteria, most notably the potentially dangerous opportunist pathogen *Pseudomonas aeruginosa,* porin function is even more restrictive. The high-flux channels of *Escherichia coli* are replaced in *Pseudomonas aeruginosa* by a low-efficiency porin, OprF, that restricts diffusion to about 1% of the rate through the channels of other Gram-negative bacteria. Only about 2% of OprF porins are in an active, open state at any one time. The absence of efficient porin channels and the low permeability of the rigid LPS of its outer membrane are major contributors to the characteristic intrinsic resistance of *Pseudomonas aeruginosa* to both hydrophilic and lipophilic agents.

Although mycobacteria are formally classified as Gram-positive, the organization of the dense lipid material in the outer layers of the cell envelope resembles the outer membrane of Gram-negative bacteria. As previously discussed, the extreme impermeability of the outer lipid coating of mycobacteria is the major determinant of the general resistance of these organisms to antibacterial agents. How therefore do the effective antituberculosis drugs (and indeed, nutrients) gain access to the bacterial cytoplasm? While slow diffusion across the lipid bilayer is probable, recent discoveries suggesting the existence of porins in the bilayer indicate the possibility of another route of access. A porin protein, MspA, with high channel activity, has been isolated from *Mycobacterium smegmatis.* Recently the MspA protein was crystallized and subjected to X-ray analysis. A homo-octameric structure was revealed with a single central channel which would readily permit the passage of isoniazid, ethambutol and pyrizinamide and small, water-soluble nutrients (Figure 7.1). It remains to be seen whether comparable structures can be defined in *Mycobacterium tuberculosis*. Further research into this possibility would be relevant to the development of new antituberculosis drugs.

It had been thought that the barrier function of the outer membrane provided an adequate explanation for the intrinsic resistance of Gram-negative bacteria to many drugs. However, it is now recognized that even a highly effective permeability barrier cannot completely stem the influx of drugs. An interesting example is provided by hydrophilic β-lactams. These compounds cross the outer membrane of *Escherichia coli* through the porin system and by slow diffusion across the lipid bilayer. The half-equilibration time of β-lactams into the periplasmic space is less than 1 s. In porin-deficient mutants, the only route of drug ingress is by diffusion across the lipid bilayer. Even in this situation, the half-equilibration time is only a few minutes. Thus although the minimal inhibitory concentrations of β-lactams against the porin-deficient cells are significantly increased, there is still effective access to the penicillin-binding proteins on the outer face of the cytoplasmic membrane.

7.2 Multidrug efflux

Because the permeability barrier of the lipid bilayer of the outer membrane cannot fully account for the intrinsic resistance of many Gram-negative bacteria to antibacterial agents, there must be some additional factor at work. In fact, wild-type strains of many bacteria, both Gram-negative and Gram-positive, extrude a wide range of antibiotics, including β-lactams, tetracyclines, chloramphenicol, macrolides and fluroquinolones as well as antiseptic agents into the external medium by drug efflux pumps. The drug efflux activities of various strains of *Pseudomonas* correlate well with their relative levels of antibiotic resistance. The absolute levels of intrinsic resistance are determined

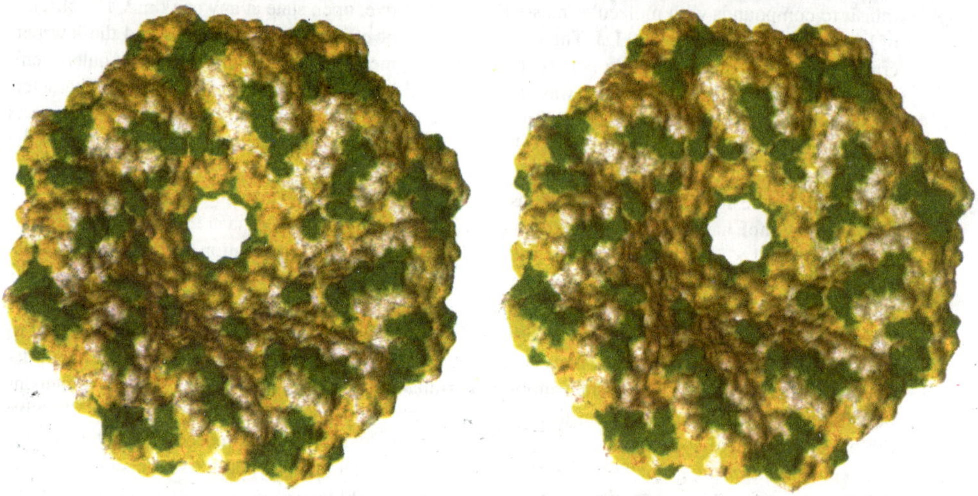

FIGURE 7.1 Stereo view from the external environment into the MspA porin channel from *Mycobacterium smegmatis*. The surfaces of polar amino acid residues are shown in green and nonpolar residues in yellow. If the photograph is held approximately 50 cm from the eyes and attention is concentrated on the space between the two images, with practice a three-dimensional image can be seen between the two outer images. The images were generated from X-ray analysis of the crystalline porin protein. [Taken with permission from M. Faller, M. Niederweis and G. E. Schulz *Science* **303**, 1189 (2004).]

by the synergism between the low permeability of the outer membrane and the drug efflux systems, which together depress the prevailing intracellular drug concentrations below effective inhibitory levels.

An extraordinary diversity of drug efflux pumps has been discovered in recent years, not only in bacteria, but also in fungi and protozoa and even in mammalian cells. The ability of living cells to reverse the influx of harmful chemicals has a long evolutionary past and is a major contributor to the intrinsic resistance of cells. The upregulation and acquisition of drug efflux systems by gene transfer in acquired drug resistance is discussed in Chapter 9. An extensive review of the full range of drug efflux pumps is beyond the scope of this book and several references are available in 'Further reading' to take the reader deeper into this complex field. However, it is evident that the many different pumps which have been characterized can be grouped into four major families:

1. the major facilitator family (MFS),
2. the resistance-nodulation-division (RND) family,
3. the small multi-drug resistance (SMR) family,
4. the ATP-binding cassette (ABC) family.

The efflux pumps are able to drive a wide range of toxic chemicals (Table 7.2) out of the cells against a concentration gradient using carrier molecules and are therefore examples of active transport. The energy for active transport is generated by the proton motive force in the MFS, RND, and SMR pumps. As its name implies, the ABC family consumes ATP to drive drug extrusion. A fifth family, multidrug and toxic compound extrusion (MATE), has recently been described in *Escherichia coli.*

MFS efflux pumps are found in both Gram-positive and Gram-negative bacteria. The QacA system (a 14-transmembrane domain protein) in *Staphylococcus aureus* extrudes antiseptic compounds such as chlorhexidine and cetyltrimethylammonium bromide. NorA, a 12-transmembrane domain protein expressed in *Staphylococcus aureus,* extrudes fluoroquinolones, chloramphenicol, antiseptics and a wide range of other compounds.

The presence of the outer membrane in Gram-negative bacteria requires the cooperation of additional proteins to ensure the export of chemicals across this barrier. The membrane fusion protein (MFP) links

TABLE 7.2 Examples of antimicrobial efflux pumps. The substrates list provides examples only and the range of substrates taken up by the pumps is often much wider

Pump type	*Organism*	*Antimicrobial substrates*
ABC		
MacAB-TolC	*Escherichia coli*	Macrolides
Abc23	*Enteroccocus faecalis*	Lincomycin
Cdr1	*Candida albicans*	Azoles
Pfmdr1	*Plasmodium falciparum*	Choroquine, artemisinin
SMR		
EMRe	*Escherichia coli*	Quaternary ammonium antiseptics,
Mmr	*Mycobacterium tuberculosis*	Erythromycin
MFS		
QacA, QacB	*Staphylococcus aureus*	Chlorhexidine, quaternary ammonium antiseptics
MdfA	*Escherichia coli*	Norfloxacin, erythromycin, quaternary ammonium antiseptics
Mdr1	*Candida albicans*	Azoles, benomyl
RND		
AcrAB/Tolc	*Escherichai coli*	β-Lactams, macrolides
MexAB/OprM	*Pseudomonas aeruginosa*	β-Lactams, macrolides, Fluoroquinolones
MexXY/OprE	*Pseudomonas aeruginosa*	Aminoglycosides

the efflux pump in the cytoplasmic membrane with the outer membrane factor (OMF) which is embedded in the outer membrane and extends through the periplasm. The whole assembly provides a conduit for the extrusion of toxic chemicals into the extracellular space. Figure 7.2 illustrates how this system may be arranged in the cell envelope; the mechanism of the link with the energy-generating proton motive force of the cytoplasmic membrane is not known.

Members of the RND family are mostly found in Gram-negative bacteria, although evidence for them has been detected in Gram-positive cells. The RND pumps are nonspecific in their uptake of chemicals and hence confer intrinsic resistance to a wide range of antibacterial agents (Table 7.2) There are 12 transcytoplasmic membrane domains (TM) together with two large extracytoplasmic domains between TM1 and TM2 and between TM7 and 8. The MFP and OMF proteins complete the assembly. The best-known RND pump is the AcrB transporter in *Escherichia coli*. Here the MFP protein is designated as AcrA and the OMF as TolC. The AcrAB complex confers intrinsic resistance to large lipophilic drugs, such as erythromycin, fusidic acid and detergents, that traverse the porin channels with difficulty. Susceptibility to smaller antibiotics, including tetracyclines, chloramphenicol and fluoroquinolones, which diffuse through the porin channels, remains high. The rate of influx of these drugs overwhelms the capacity of the AcrAB system to maintain cytoplasmic concentrations below inhibitory levels. Nevertheless, deleterious mutations affecting AcrAB greatly enhance the sensitivity of *Escherichia coli* to a

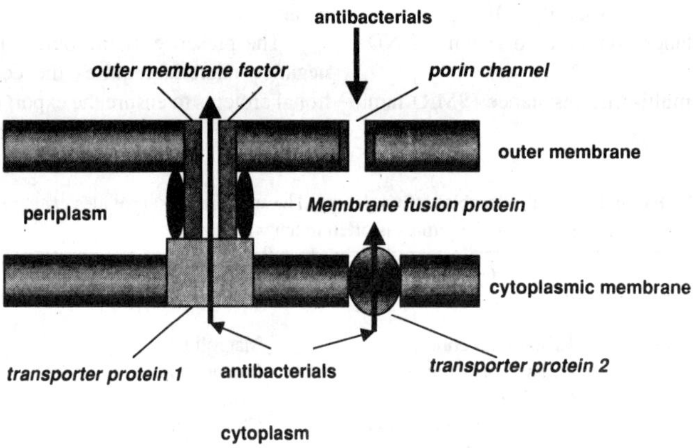

FIGURE 7.2 Suggested arrangement of the components of bacterial multidrug efflux pumps in Gram-negative bacteria. RND and MFS pumps (transporter protein 1) use accessory proteins in Gram-negative bacteria as channels or pores to direct antibiotics out of the cells into the external environment. SMR and ABC pumps (transporter protein 2) extrude antibiotics across the cytoplasmic membrane. In Gram-positive bacteria, the extruded antibiotics then diffuse directly through the peptidoglycan cell wall to the cell exterior. The pumps also capture antibiotics in the periplasm of Gram-negative bacteria. In Gram-negative bacteria, antibiotics delivered by the SMR and ABC pumps may diffuse outwards via porin channels. RND, MFS and SMR pumps are energized by the proton motive force across the cytoplasmic membrane: ABC pumps are driven by the hydrolysis of ATP. The mechanism of coupling energy consumption to pump function is at present not known. The arrows indicate the direction of antibiotic fluxes.

variety of drugs. The AcrAB-TolC system of *Escherichia coli* is homologous with the MexAB-OprM pump of *Pseudomonas aeruginosa* and Acr in *Salmonella* spp. and *Haemophilus influenzae*. Mutational inactivation of either MexA or OprM in wild-type *Pseudomonas aeruginosa* results in a marked increase in cellular sensitivity to many antibacterial agents.

There is evidence for SMR efflux pumps in both Gram-negative and Gram-positive bacteria and in *Mycobacterium tuberculosis*. The SMR pump, EmrE, in *Escherichia coli* is capable of transporting a diverse range of antibacterial agents (Table 7.2), but the details of its organization are not clear.

The complete sequence of the genome of the yeast *Saccharomyces cerevisiae* predicts that it contains 30 genes for ABC pumps and 28 genes for MFS pumps, although to what extent these genes are constitutively expressed and affect intrinsic drug resistance is largely unknown. The role of ABC and MFS pumps in acquired drug resistance in eukaryotic cells is reviewed in Chapter 9. At least two ABC pumps have been identified in *Escherichia coli*: arcAB and MacAB-TolC, the latter contributing to the intrinsic resistance to macrolide antibiotics. The bacterial pathogen *Enterococcus faecalis* also expresses four of these pumps that confer substantial intrinsic resistance to the lincosamide antibiotics.

An extraordinary property common to all the efflux pumps is their broad substrate specificity, enabling the extrusion of a diverse range of antimicrobial compounds. Until recently, the structural basis for this broad substrate specificity remained elusive. However, the solution of the crystal structure of the AcrB pump from *Escherichia coli* has now provided a remarkable insight into this puzzle. The transmembrane protein was crystallized both in isolation and complexed with four structurally dissimilar compounds, including the fluoroquinolone ciprofloxacin, the quaternary antiseptic dequalinium, and rhodamine. AcrB is a homotrimer, which in addition to threading through the cytoplasmic membrane, forms a large cavity facing the cytoplasm. This cavity contains the ligand-binding domain. The ligands bind to various positions in the cavity using different subsets of hydrophobic amino acids lining the cavity. The binding interactions are therefore believed to be mainly hydrophobic, perhaps also involving aromatic π-π interactions. The large size of the central cavity and the predominance of hydrophobic binding interactions with drug molecules largely explains the broad range of antibacterial agents that the AcrB pump extrudes from the bacterial cytoplasm. It seems likely that similar ligand-binding motifs may underly the broad specificity of many other drug efflux pumps.

The upregulation of drug efflux systems in acquired drug resistance is considered in greater detail in Chapter 8. It is important to realize, however, that the contribution of constitutively expressed multidrug efflux pumps in wild-type bacteria to intrinsic drug resistance depends on the rate of the inward movement of drugs across the cell envelope. If the barrier function is breached by mutational changes or treatment with EDTA, which chelates divalent cations and thereby disrupts the LPS of the outer membrane of Gram-negative bacteria, the efflux systems may be overwhelmed by the inward rush of drug molecules, and intrinsic resistance will be lost. The influence of the various components of bacterial cell envelopes on the uptake of drugs is summarized in Table 7.3.

7.3 Facilitated uptake of antimicrobial drugs

7.3.1 Antibacterial agents

As we have seen, the inward diffusion of antibacterial agents across the outer membrane of Gram-negative bacteria is essentially a passive process. Hydrophilic compounds move through the water-filled porin channels, whereas lipophilic agents diffuse through the lipid bilayer. Strongly charged cationic agents such as quaternary ammonium antiseptics and chlorhexidine, and polycationic antibiotics such as the polymyxins and aminoglycosides, destabilize and disorganize the outer membrane and thereby promote their own access to the inner cellular layers. As we saw in Chapter 2, antiseptics and polymyxins also disrupt the barrier function of the cytoplasmic membrane and finally penetrate the cytoplasm. Lipophilic agents, including sulfonamides, rifamcyins and macrolides, diffuse passively across the cytoplasmic membranes of Gram-negative and Gram-positive bacteria. Hydrophilic compounds, however, are unlikely to achieve inhibitory intracellular concentrations unless there is

TABLE 7.3 Features of the bacterial cell envelope that influence the uptake of antibacterial agents

Envelope structure	*Effects on drug uptake*
Lipid bilayer of Gram-negative outer membrane	Retards diffusion of both water-soluble and lipophilic compounds
High-efficiency porin channels of Gram-negative bacteria	Facilitate diffusion of water-soluble molecules of molecular mass up to ~600 Da
Low-efficiency porin of *Pseudomonas aeruginosa*	Permits only slow diffusion of water-soluble antibiotics
Teichoic acids of Gram-positive bacteria	Strongly anionic character could in principle affect uptake of ionized molecules
Lipid bilayer of cytoplasmic membrane	Rates of passive diffusion depend on lipophilicity of permeant. Permits little or no passive diffusion of water-soluble or strongly ionized molecules
Facilitated transport systems of cytoplasmic membrane	Markedly enhance rates of transfer of nutrients and structural analogues, e.g. D-cycloserine, fosfomycin. Energy coupling leads to accumulation against concentration gradients
Multidrug efflux systems	Reduce intracellular drug concentrations by effluxing into external medium. Occur in both wild-type and drug-resistant mutant bacteria

some form of active or facilitated transport or intracellular sequestration of unbound drug to maintain a steep inward concentration gradient. Examples of facilitated drug transport into the bacterial cytoplasm are described below.

D-Cycloserine

This antibiotic, a structural analogue of D-alanine, is transported across bacterial cytoplasmic membranes by alanine permeases. D-Cycloserine enters *Streptococcus faecalis* using this transport system and competitively inhibits the uptake of both stereoisomers of alanine. In *Escherichia coli,* D- and L-alanine are transported by separate isostere-specific permeases and D-cycloserine uses only the D-isomer-specific system. There are high- and low-affinity transport systems for D-alanine, and at minimal antibacterial concentrations (~4 μM) D-cycloserine favors the high-affinity carrier. The high-affinity D-alanine permease is energy coupled to the proton motive force, ensuring accumulation of D-alanine and D-cycloserine against their concentration gradients. The accumulated D-cycloserine effectively inhibits the intracellular target enzymes, L-alanine racemase and D-alanyl-D-alanine synthetase, involved in peptidoglycan biosynthesis (Chapter 2).

Fosfomycin (phosphonomycin)

This simple phosphorus-containing antibiotic uses two different physiological transport systems to gain access to the bacterial cytoplasm:

1. The structural resemblance of fosfomycin to α-glycerophosphate enables it to use the permease for this important biochemical. As expected, the uptake of fosfomycin is competitively inhibited by high concentrations of α-glycerophosphate.
2. The permease for hexose 6-phosphates in certain enterobacteria and staphylococci is also exploited by fosfomycin.

Both transport systems are induced by their normal permeants, although not by fosfomycin. The therapeutic efficacy of fosfomycin in experimental infections can be enhanced by pretreatment of the infected animals with glucose 6-phosphate, which induces the bacterial permease, resulting in a higher intracellular concentration of fosfomycin. In Gram-negative bacteria, fosfomycin crosses the outer membrane via nutrient channels used by α-glycerophosphate and hexose 6-phosphates.

Tetracyclines

Tetracyclines cross the outer membrane of Gram-negative bacteria through the OmpC and OmpF porin channels, probably as a chelation complex with a magnesium ion. Mutant cells with decreased OmpF expression exhibit some resistance to tetracycline, although not to the more lipophilic derivative, minocycline, which is thought to diffuse in its uncomplexed form across the lipid bilayer rather than through the porin channels.

It has long been known that tetracyclines are accumulated against their concentration gradients by Gram-negative and Gram-positive bacteria. This active process, which is energized by the proton motive force across the cytoplasmic membrane, partially explains the antibacterial specificity of tetracyclines. The energy-dependent accumulation of tetracyclines against a concentration gradient appears to have some of the hallmarks of facilitated diffusion mediated by a carrier system associated with the cytoplasmic membrane. However, no such carrier for the inward transport of tetracycline across the cytoplasmic membrane has ever been identified. Furthermore, there is scant evidence that the rate of tetracycline influx is saturable at high concentrations of the drug—a defining characteristic of carrier-mediated facilitated transport. How, then, is the energy-dependent accumulation of tetracyclines to be explained? The answer probably is that it is at least partly due to the physicochemical properties of the tetracycline molecule.

The tetracycline-Mg^{2+} chelate is positively charged and probably accumulates in the bacterial periplasm of Gram-negative cells, owing to the Donnan potential across the outer membrane. Reversible dissociation of the chelate releases the lipophilic, uncharged tetracycline molecules. Calculations based on the use of microscopic dissociation constants, which define the protonation of tetracycline, show that approximately 7% of the molecules are uncharged at physiological pH. The uncharged molecules, unlike their charged counterparts, diffuse rapidly across the cytoplasmic membrane into the cytoplasm. The proton gradient across the cytoplasmic membrane maintains the internal pH about 1.7 pH units higher than the pH of the external medium. The effect of the higher internal pH is to significantly increase the fraction of negatively charged tetracycline molecules within the cells. When equilibrium is reached by passive diffusion, the concentration of uncharged molecules must be the same on both sides of the membrane. The total concentration of tetracycline, i.e. uncharged plus charged molecules, is therefore higher in the bacterial cytoplasm than in the medium. Calculations based upon these assumptions predict that the intracellular concentration of tetracycline in normally metabolizing bacterial cells should be approximately four times that in the external medium. Direct measurements in *Escherichia coli*, however, reveal a 15-fold difference between the internal and external concentrations. The higher internal concentration of magnesium ions may also contribute to the sequestration of tetracycline. Up to 30% of the intracellular tetracycline is bound to the 30S subunit of the ribosomes, probably as an Mg^{2+} complex (Chapter 5). The maintenance of the pH gradient across the membrane depends on the energy metabolism of the cell. Compounds that collapse the gradient directly or indirectly by inhibiting energy metabolism, prevent the intracellular accumulation of tetracycline and promote the release of previously accumulated drug into the external medium.

The model described here provides a probable explanation for the energy-dependent accumulation of tetracycline by bacterial cells, at least in Gram-negative organisms. However, before the possibility of carrier-mediated influx is rejected entirely, it should be noted that tetracycline-specific, carrier-mediated efflux systems occur widely in tetracycline-resistant bacteria (Chapter 9). Furthermore, even in wild-type *Escherichia coli* there is evidence of a low-efficiency tetracycline-specific efflux system, probably involving a carrier system. The existence of these tetracycline efflux pumps therefore raises the possibility that tetracycline influx may also be carrier mediated.

Quinolones

Antibacterial quinolones, such as ciprofloxacin, have two ionizable centers—the carboxyl group, pK_a 7.5, and the 'distal' nitrogen of the piperazine ring, pK_a 6.5. Calculations similar to those applied to the tetracyclines show that at pH 7.4 approximately 10% of the molecular population is in the uncharged form. The precise values of these parameters for individual compounds depend on the nature of the substituents on the quinolone ring system.

Quinolone entry through the outer membranes of Gram-negative bacteria occurs mainly as charged molecules through the porin channels. Ionization of the carboxyl group enables quinolones to chelate with Mg^{2+} ions, and a substantial part of the influx through the porin channels is probably in the form of the magnesium complex. The chelates are likely to dissociate in the more acid environment of the periplasm, leading to the establishment of a Donnan equilibrium across the outer membrane, with a higher total drug concentration in the periplasm than in the external medium. This effect could explain why some quinolones are more effective against Gram-negative than against Gram-positive bacteria, possibly because of the absence of a defined periplasm in the latter organisms.

Bacterial cells accumulate quinolones and it was thought that this was an example of active drug uptake into the cytoplasm. However, much of the drug associated with Gram-negative bacteria is probably bound to surface components via magnesium chelation. Consideration of the ionization equilibria of quinolones suggests that their cytoplasmic concentration may actually be less than that in the external medium, owing to the pH gradient across the cytoplasmic membrane. When equilibrium of the uncharged molecules is reached by passive diffusion across the cytoplasmic membrane, the total internal concentration (uncharged plus charged molecules) is calculated to be less than the external concentration. It is interesting that carbonyl cyanide *m*-chlorophenylhydrazone, an agent used experimentally to collapse the proton gradient across cell membranes in bacteria, actually increases the uptake of certain quinolones into the cytoplasm. Abolition of the pH difference between the cytoplasm and the exterior is believed to establish a new equilibrium position by promoting a transient-increased influx of quinolone that equalizes the total drug concentrations on each side of the membrane. In summary, there is no definitive evidence for carrier-mediated accumulation of quinolones by bacteria, and the uptake phenomena observed can be explained largely by the physicochemical properties of the compounds and compartmental differences in pH within bacterial cells.

Aminoglycosides

These polycationic water-soluble molecules approach the molecular mass exclusion limit of 600 Da for porin-mediated transport through the Gram-negative outer membrane. There is uncertainty, therefore, as to the contribution of this mode of transport to the uptake of aminoglycosides by Gram-negative bacteria. As mentioned previously, aminoglycosides promote their own penetration by competitive displacement of the stabilizing Mg^{2+} and Ca^{2+} ions from the LPS. The large molecular size of the aminoglycoside cations in comparison with the metal cations probably causes the subsequent disorganization and disruption of the barrier function of the outer membrane. The initial uptake of dihydrostreptomycin by *Escherichia coli* is characterized by an electrostatic interaction between the positively charged guanidino centres of the antibiotic and the anionic groups of LPS. There may be some accumulation in the periplasm, followed by a slow, energy-dependent penetration into the cytoplasm. After 15–30 min, a third phase of rapid, energy-dependent intracellular accumulation of dihydrostreptomycin begins. This final phase appears to be irreversible in *Escherichia coli*, and the antibiotic can only be released from the cells by damaging the cell membranes with organic solvents such as toluene. The molecular mechanisms involved in the energy-dependent phases of aminoglycoside uptake are obscure. Although evidence for an uptake carrier system across the cytoplasmic membrane for aminoglycosides is lacking, the recent discovery of aminoglycoside-specific efflux pumps in several species of bacteria (Table 7.2) suggests that carrier-dependent influx could also exist, perhaps mediated by a physiological nutrient carrier system. The binding of aminoglycosides to 70S ribosomes enhances accumulation within bacteria. The intrinsic resistance of anaerobic bacteria to the aminoglycosides may be due to their limited ability to accumulate these antibiotics.

7.3.2 Uptake of antimicrobial drugs by eukaryotic pathogens

The mechanisms involved in the transport of drugs into fungal and protozoal pathogens have received less attention than their counterparts in bacteria. In fungi and protozoa, the cytoplasmic membranes are likely to be the major permeability barriers against drug influx. The complex outer walls of fungi could conceivably hinder the access of some larger molecules, but in general the coarseness of the chitin and mannan meshworks probably offers little resistance to drug influx.

5-Fluorocytosine

This antifungal drug provides an example of facilitated transport into fungal cells. As a close analogue of cytosine, 5-fluorocytosine is transported across the cytoplasmic membrane by the cytosine permease. Rapid intracellular metabolism of 5-fluorocytosine to several toxic pyrimidine nucleotides contributes to the maintenance of a downward concentration gradient of unchanged drug into fungal cells. A disadvantage of reliance on cytosine permease for the uptake of 5-fluorocytosine is that mutational inactivation of the transport system results in resistance to the drug.

Polyoxins and nikkomycins

The inhibitory activities of these peptidonucleoside antibiotics depend on their transport into fungal cells by a peptide permease system normally intended for the accumulation of nutrient dipeptides. However, antibiotic transport on the permease system is subject to competitive inhibition by dipeptides, which commonly occur in the blood and tissues of the infected host. As in the case of 5-fluorocytosine, resistance to polyoxins and nikkomycins also arises readily from mutations that inactivate the permease system.

Azoles

The lipophilic character of the azole antifungal drugs probably ensures their penetration of cytoplasmic membranes by passive diffusion.

Antiprotozoal drugs

With the exception of the polycationic, antitrypanosomal drug suramin, which enters the parasite complexed with serum proteins by a process of endocytosis, most antiprotozoal drugs probably diffuse passively across the cytoplasmic membranes of their target pathogens. Positively charged antimalarial compounds, such as chloroquine, subsequently accumulate within the acidic environment of the digestive vacuoles of the parasites. The binding of certain antimalarial drugs to heme released by the digestion of hemoglobin, and to the heme crystalline polymer hemozoin (Chapter 6) also contributes to the persistence of favourable chemical gradients into the vacuoles.

The antitrypanosomal drug eflornithine may enter trypanosomes by a permease that normally facilitates the uptake of ornithine and other polyamines. However, as yet there is no experimental evidence for this.

Further reading

Balkis, M. M. *et al.* (2002). Mechanisms of fungal resistance. *Drugs* **62**, 1025.

Denyer, S. P. and Maillard, J.-Y., (2002). Cellular impermeability and uptake of biocides and antibiotics in Gram-negative bacteria. *J. Appl. Microbiol.* **92**, 35S.

Faller, M. *et al.* (2004). The structure of a mycobacterial outer-membrane channel. *Science* **303**, 1189.

Hancock, R. E. W. (1997). The bacterial outer membrane as a drug barrier. *Trends Microbiol.* **5**, 37.

Lambert, P. A. (2002). Cellular impermeability and uptake of biocides and antibiotics in Gram-positive bacteria and mycobacteria *J. Appl. Microbiol.* **92**, 46S.

Mao, W. *et al.* (2001). MexXY-Opr efflux pump is required for antagonism of aminoglycosides by divalent cations in *Pseudomonas aeruginosa*. *Antimicrob. Agents and Chemother.* **45**, 2001.

McKeegan, K. S. *et al.* (2002). Microbial and viral drug resistance mechanisms. *Trends Microbiol.* **10** Suppl., S8.

Niederweis. M. (2003). Mycobacterial porins—new channel proteins in unique outer membranes. *Molec. Microbiol.* **49**, 1167.

Nikaido, H. (2003). Molecular basis of bacterial outer membrane permeability revisited. *Microbiol. Molec. Biol. Rev.* **67**, 593.

Schulz. G. (2002). The structure of bacterial outer membrane proteins. *Biochem. Biophys. Acta* **1565**, 308.

Tute, M. S. (1972). Principles and practice of Hansch analysis: a guide to the structure–activity relationships for the medicinal chemist. *Adv. Drug. Res.* **6**, 1.

Yu, E. W. *et al.* (2003) . Structural basis of multiple drug-binding capacity of the AcrB multidrug efflux pump. *Science* **300**, 976.

Zgurskaya, H. I. (2002). Molecular analysis of efflux pump-based antibiotic resistance. *Int. J. Med. Microbiol.* **292**, 95.

Chapter eight

The genetic basis of resistance to antimicrobial drugs

The development of safe, effective antimicrobial drugs has revolutionized medicine in the past 70 years. Morbidity and mortality from microbial disease have been drastically reduced by modern chemotherapy. Unfortunately, micro-organisms are nothing if not versatile, and the brilliance of the chemotherapeutic achievement has been dimmed by the emergence of microbial strains presenting a formidable array of defences against our most valuable drugs. This should not surprise us, since the evolutionary history of living organisms demonstrates their adaptation to the environment. The adaptation of micro-organisms to the toxic hazards of antimicrobial drugs is therefore probably inevitable. In Chapter 7 we saw that in many species of bacteria a degree of intrinsic resistance to toxic chemicals is conferred by cellular permeability barriers and low levels of expression of a range of multidrug efflux pumps. Both elements of defence can be enhanced in acquired drug resistance by genetic changes in response to higher levels of drug challenge. The extraordinary speed with which antibiotic resistance has spread among bacteria and certain viruses such as HIV during the era of chemotherapy is due, in large measure, to the remarkable genetic flexibility of these organisms. Figure 8.1 provides an example in the alarming rise of a drug-resistant pathogen, methicillin (or multidrug)-resistant *Staphylococcus aureus*, in part of the United Kingdom in just nine years.

The first account of microbial drug resistance was given by Paul Ehrlich in 1907, when he encountered the problem soon after the development of arsenical chemotherapy against trypanosomiasis. As the sulfonamides and antibiotics were brought into medical and veterinary practice, resistance against these agents began to emerge. Resistance to antibacterial and antimalarial drugs is now widespread, and increasing resistance to antifungal and antiviral drugs is also a major concern. Our intention in this chapter is to give an outline of the genetic background to the problem of drug resistance; in Chapter 9 we describe the major biochemical mechanisms that give rise to resistance.

The tremendous advances made in the science of bacterial genetics over the past 60 years have found a most important practical application in furthering our understanding of the problem of drug resistance. As a result, we now have a fairly complete picture of the genetic factors underlying the emergence of drug-resistant bacterial populations. Although the study of the genetics of resistance in pathogenic fungi and protozoa is less developed, advances in DNA sequencing technology should bring about significant improvements in our understanding of these organisms. The depressingly rapid emergence of drug-resistant variants of the human immunodeficiency virus during the chemotherapy of AIDS has given a powerful impetus to the study

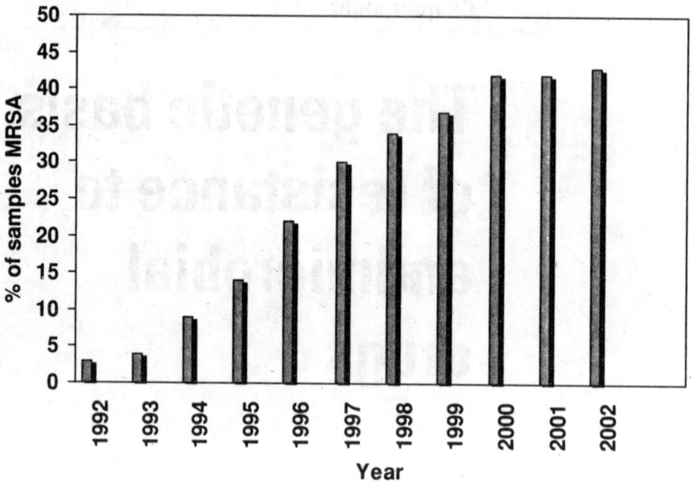

FIGURE 8.1 The recent rise in the incidence of multidrug-resistant *Staphylococcus aureus* in England and Wales. The vertical axis indicates MRSA as a percentage of all samples of *Staphylococcus aureus* examined in clinical laboratories. (Source: Health Protection Agency.)

of the genetic and biochemical basis of resistance to antiviral drugs.

The early studies on the genetics of drug resistance were bedeviled by an exhausting controversy. On the one hand were those who believed that the development of a resistant cell population could be explained by the phenotypic adaptation of the cells to an inhibitory compound without significant modification in their genotype. The opposing faction took the view that any large population of cells which was sensitive overall to a drug was likely to contain a few genotypically resistant cells. The continued presence of the drug resulted in the expansion of the numbers of resistant cells by a process of selection.

Evidence gathered over the years strongly supports the second of these two theories. As we shall see, there are examples of phenotypic adaptation to antimicrobial drugs, but such cells are usually genotypically different from wild-type cells. When the selective pressure applied by an antimicrobial drug is removed, the resistant microbial population may revert to drug sensitivity if the resistant cells are at a selective disadvantage compared with drug-sensitive cells in a drug-free environment and could therefore eventually be outnumbered by the sensitive cells.

8.1 Mutations and the origins of drug-resistance genes

Once it was accepted that drug-resistant organisms are genetically different from the wild types, it was natural to consider how such differences might arise. One obvious possibility is that of spontaneous mutations. These can arise in several ways:

1. Damage to the genome caused by adverse environmental factors, including ionizing radiation and chemical mutagens.
2. Base-pairing errors during genomic replication.
3. Frameshift mutations caused by the deletion of segments of DNA which frequently occur at short DNA repeat sequences.
4. Frameshift mutations caused by the intragenic insertion of mobilizable genetic material, such as transposons, which corrupt the correct flow of information from the wild-type genome.

Spontaneous mutations are relatively rare—on the order of one mutation per 10^7–10^{11} cells per generation—although in organisms lacking a proofreading mechanism during genomic replication, as in HIV, the mutation rate is much higher. There is also a mutator phenotype in some bacteria which ensures a much higher mutation frequency than normal. An interesting example is that of *Helicobacter pylori*, the bacterium associated with peptic ulcer disease and gastric cancers. As many as 33% of *Helicobacter pylori* strains show an abnormally high rate of mutation to antibiotic resistance. The nature of this high mutation frequency is not known at present, but it is of potential relevance to the clinical challenge of eliminating *Helicobacter pylori* from the stomach by treatment with antibiotics. When the vast numbers of organisms in microbial populations are considered, the probability of even low mutation rates causing drug resistance is quite high. The simple and elegant technique of replica plating convincingly demonstrates that spontaneous mutations to drug resistance can occur in drug-sensitive bacterial populations in the absence of drugs (Figure 8.2). A spontaneous mutation may occasionally cause a large increase in resistance, but resistance often develops as a result of numerous mutations, each giving rise to a small increase in resistance. In this situation, highly resistant organisms emerge only after prolonged or repeated exposure of the microbial population to the drug.

8.1.1 Spontaneous mutations and drug resistance in HIV

A major challenge to the effective treatment of AIDS is the unique and alarming speed with which HIV becomes resistant to every drug deployed against it, including inhibitors of viral reverse transcriptase and HIV protease. The origins of drug resistance in HIV lie in the high rate of viral replication and the ease with which spontaneous mutations arise in its RNA genome. As a single-stranded RNA virus, HIV lacks a proofreading mechanism to eliminate sequence errors resulting from the low fidelity of HIV reverse transcriptase. As a result, mutations occur with high frequency. During the course of an infection, the combination of high replication and mutation rates permits rapid and extensive evolution of the viral population in response to immunological and chemotherapeutic challenges to its survival. For example, within weeks of starting treatment with the reverse transcriptase inhibitor lamivudine, spontaneous mutation results in the replacement in the reverse transcriptase of the circulating viruses of methionine-184 by valine, a change associated with high-level resistance to lamivudine.

Resistance to some other drugs, such as azidothymidine, develop through successive mutations which progressively reduce the drug sensitivity of the target enzyme. The loss of sensitivity to an inhibitor can be associated with reduced catalytic efficiency, which places the virus particles at a competitive disadvantage compared with viruses with unimpaired enzyme. However, the reduction in enzymic efficiency may be compensated by further mutations which progressively restore enzymic activity. A recent alarming discovery is that under laboratory conditions, AZT and lamivudine adversely affect the fidelity of reverse transcriptase, thus further increasing the frequency of mutations. The clinical significance of this finding has yet to be explored. As discussed in Chapter 4, the clinical approach to coping with the rapid acquisition of drug resistance by HIV is to treat patients with a combination of drugs. In this way, the emergence of resistant viruses can be delayed by months or even years. Even so, the eventual emergence of viral populations resistant to multidrug therapy may be inevitable.

8.1.2 Origin of clinically important resistance genes in bacteria

Originally it was believed that spontaneous mutations followed by the selection of resistant organisms in the presence of a drug explained the emergence of drug-resistant populations. However, while this appears to be true in the case of HIV, the discovery that bacteria can acquire additional genetic material by conjugation, transformation and transduction led to the conclusion that spontaneous mutations make an important but not exlusive contribution to the emergence and spread of drug resistance in bacteria. Mutations underlie the upregulation of drug efflux pumps and the reduction or loss of porin function in many bacteria as well as the increased expression of constitutive β-lactamases in pathogens such as *Enterobacter cloacae* and *Citrobacter freundii*. Spontaneous mutations also lead to the progressive modification of β-lactamases, enabling

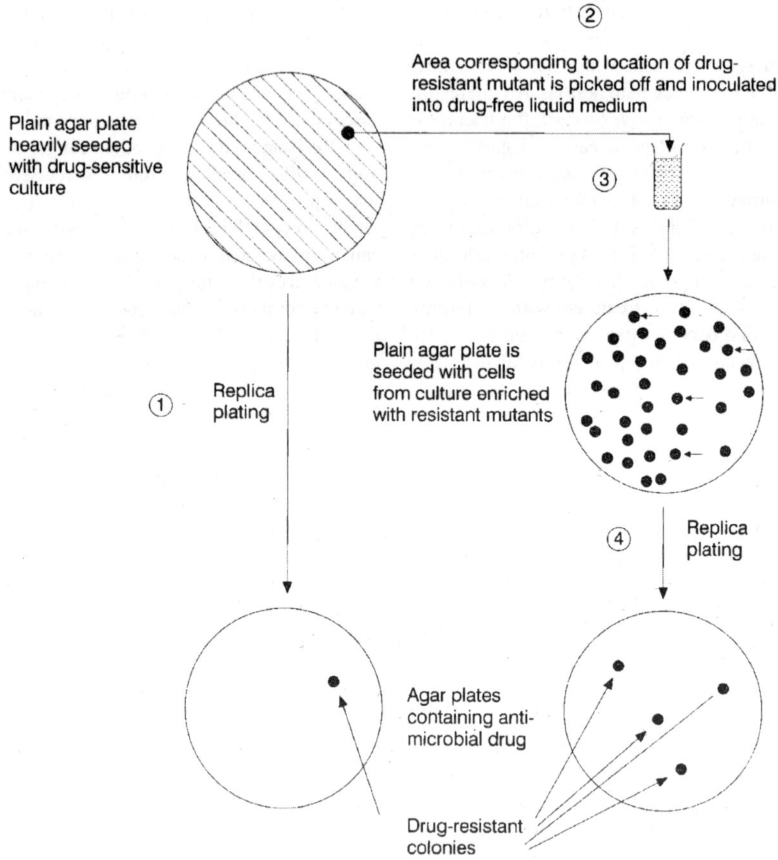

FIGURE 8.2 The technique of replica plating reveals the existence of drug-resistant cells in a population that is overall drug-sensitive. A plain agar plate is heavily seeded with cells from the drug-sensitive culture and incubated until growth occurs. Cells are transferred by a velvet pad to a plate containing the antibacterial drug; this plate is then incubated and the position of any colonies noted. The area on the drug-free plate corresponding to the location of the resistant colony on the drug plate is picked off and cultured in drug-free medium. Although still contaminated with sensitive cells, this culture will contain many more resistant cells than the original culture. Plating out of the 'enriched' culture on a plain plate followed by replication to a drug plate therefore reveals a higher number of drug-resistant colonies. The experiment shows that drug-resistant mutants occur in a bacterial population not previously exposed to the drug.

them to cope with the many novel chemical variants of β-lactams. Strains of *Mycobacterium tuberculosis* resistant to isoniazid, rifampin, pyrizinamide, ethambutol and streptomycin can all be explained by mutations in the genes encoding the target sites for these drugs. Resistance to the quinolones regularly arises through mutations in genes encoding the target DNA gyrase enzyme.

As we shall see in the next chapter, the biochemical machinery conferring bacterial resistance to drugs of major importance in medicine can be complex. Understandably, therefore, there is considerable interest in the origins of the genes that variously encode drug-inactivating enzymes, drug efflux pumps, enzymes that depress drug sensitivity by the covalent modification of drug targets and proteins that block the binding of drugs to their targets. Antibiotic-producing bacteria, such as the streptomycetes, protect themselves against the toxic effects of their own antibiotics with enzymes that inactivate aminoglycosides, chloramphenicol and β-lactams. In addition, many streptomycetes express β-lactamases even though they do not produce β-lactams, presumably as a protective measure against β-lactams synthesized by other organisms in the micro-environment. Genes encoding the pumped efflux of tetracyclines, proteins that protect ribosomes against tetracyclines, and the enzymic modification of ribosomal RNA associated with resistance to erythromycin, have all been identified in streptomycetes. Nucleic acid and protein sequence data support the suggestion that genes for aminoglycoside-inactivating enzymes found in aminoglycoside-resistant clinical isolates may have originated from streptomycetes.

Mosaic genes

Although bacterial genes which encode antibiotic-inactivating enzymes and drug efflux pumps almost certainly evolved in the very distant past, antibacterial drug resistance mediated by mutations is generally thought to have emerged during the modern era of chemotherapy. In addition to point mutations, deletions and insertions, there is also the remarkable phenomenon of mosaic genes which arise by interspecies genetic recombination. By far the most common mechanism of resistance to β-lactam antibiotics is that of antibiotic hydrolysis by β-lactamases, which are probably of ancient origin. However, β-lactam resistance in several important pathogens, including *Haemophilus influenzae*, *Neisseria gonorrhoeae*, *Streptococcus pneumoniae*, *Staphylococcus aureus* and *Staphylococcus epidermidis*, can also be caused by penicillin-binding proteins (Chapter 2) with reduced affinity for β-lactams. This type of resistance is relatively rare because the killing action of β-lactams depends on drug interactions with several high-molecular weight PBPs, and resistance therefore necessitates reductions in β-lactam affinity in each PBP. Although it is conceivable that such reductions in affinity could have arisen gradually from incremental changes in protein structure that were due to the accumulation of mutations in the PBP genes, it is clear that recombination amongst PBP genes from different species is a major cause of the low-affinity PBP phenotype in bacteria.

Analysis of the sequences of genes for PBP2 from penicillin-sensitive and penicillin-resistant meningococci and gonococci reveals that whereas the sequences from penicillin-sensitive bacteria are uniform, the resistant gene sequences have a mosaic structure. The mosaics are created when regions essentially identical with those from penicillin-sensitive bacteria recombine with regions that have significantly divergent sequences. The mosaic genes encode PBP2 variants with decreased affinity for penicillin. Sequence information obtained from bacterial DNA databases show that the divergent regions in the mosaic genes originate from *Neisseria flavescens* and *Neisseria cinerea*. PBP2 prepared from specimens of *Neisseria flavescens* preserved from the preantibiotic era has a much lower affinity for penicillin than PBP2 from either *Neisseria gonorrhoeae* or *Neisseria meningitidis*. The mosaic genes are thought to have arisen by interspecies recombination following transformation by DNA released from lysed cells (see later discussion) amongst these bacteria. Mosaic genes encoding low-affinity PBPs 1a, 2x, 2b and 2a have been isolated from *Streptococcus pneumoniae* resistant to both penicillins and cephalosporins. Pneumococci are also readily transformable, and the divergent regions of the mosaic genes appear to have originated from several other bacterial species. However, not all low-affinity PBPs are the result of mosaic gene formation. An important example is that of PBP2a, which is encoded by the *mecA* gene responsible for the notorious methicillin-resistant *Staphylococcus aureus*. The *mecA* gene is located on a large (~50 kilobases) DNA element inserted into the bacterial chromosome. This so-called resistance island encodes proteins that are homologous with transposases and integrases (see later discussion) which probably catalyze both excision and integration of the *mecA* gene. *Staphylococcus aureus* may have acquired *mecA* by gene transfer from another organism.

Interspecies recombination amongst meningococci also resulted in mosaic genes that encode sulfonamide-resistant dihydropteroate synthase (Chapter 4). Allelic variations in the *tetM* gene, which determines the ribosome protection form of resistance to tetracycline (Chapter 9), are due to recombination amongst the distinct *tetM* alleles found in *Staphylococcus aureus* and *Streptococcus pneumoniae*. *TetM* genes are widely distrbuted in both Gram-positive and Gram-negative bacteria. Finally, it should be noted that the generation of mosaic genes by interspecies recombination in bacteria is not limited to resistance genes. The phenomenon is widespread in bacteria and underlies, for example, the highly divergent genes that encode the proteins of the outer membranes of *Neisseria* spp.

8.2 Gene mobility and transfer in bacterial drug resistance

The spread of drug resistance amongst bacterial pathogens owes much to the remarkable ability of bacteria to mobilize genes in both chromosomal and plasmid DNA and to transfer and exchange genetic information. Evidence that drug resistance could be transferred from resistant to sensitive bacteria came from combined epidemiological and bacterial genetic studies many years ago in Japan. The first clue was provided by the isolation, from patients suffering from dysentery, of strains of shigella resistant to several drugs, including sulfonamides, streptomycin, chloramphenicol and tetracycline. Even more striking was the discovery that both sensitive and multiresistant strains of shigella could occasionally be isolated from the same patient during the same epidemic. Most patients harbouring multiresistant shigella also had multiresistant *Escherichia coli* in their intestinal tracts. This suggested that drug resistance markers might be transferred from *Escherichia coli* to shigella and vice versa. Subsequently it was confirmed that Gram-negative bacteria can indeed transfer drug resistance not only to cells of the same species but also to bacteria of different species and genera. As we shall see, the phenomenon of horizontal gene transfer, as it is now termed, is not confined to Gram-negative bacteria but also occurs in Gram-positive organisms. However, before we describe the transfer of drug-resistance genes between bacterial cells, we must first consider the movement of genes within the bacterial genome itself.

8.2.1 Transposons and integrons

For many years the movement of genes among plasmids and chromosomes was believed to result from classic recombination dependent on the product of the bacterial *recA* gene and the reciprocal exchange of DNA in regions of considerable genetic homology. This permits the exchange of genetic information only between closely related genomes. However, such a restricted phenomenon seemed unlikely to explain the widespread distribution of specific resistance determinants. It is now clear that the acquisition of genetic material by plasmids and chromosomes in both Gram-negative and Gram-positive bacteria is not limited by classic *recA*-dependent recombination. Replicons, known as transposons, can insert themselves into a variety of genomic sites that often have little or no homology with the inserting sequence, although such transposition events are rare, one in 10^5–10^7 cells per generation. Because there are many possible transposon, insertion sites in the bacterial genome, a higher frequency of insertion would probably result in too great a rate of gene disruption and mutation. In the simplest transposons, the whole of the genetic information is concerned with the insertion function. Insertion sequences (IS elements) are sequences of approximately 750–1600 base pairs encoding a specific endonuclease called a transposase. The IS elements are flanked by inverted repeats of 15–20 base pairs that are characteristic of individual transposons. Immediately adjacent to the inverted repeats are short direct repeats (5–11 base pairs) whose sequences depend on the target site where the transposon is inserted.

The genes for drug resistance are carried by composite transposons designated by the prefix Tn. In the class 1 transposon, Tn9, the gene encoding the enzyme that confers resistance to chloramphenicol, chloramphenicol acetyl transferase (Chapter 9), is flanked by two IS elements. These genes are again bounded by inverted repeats which in turn are flanked by short direct repeats (Figure 8.3a). Tn3 (Figure 8.3b) is a class 2, complex transposon which contains the genes for the transposase and for resolvase, an enzyme that catalyzes recombination between the insertion sequences. These genes, together with the gene for β-lactamase, are flanked by the inverted repeats. Transposons carrying arrays of drug-resistance genes have been

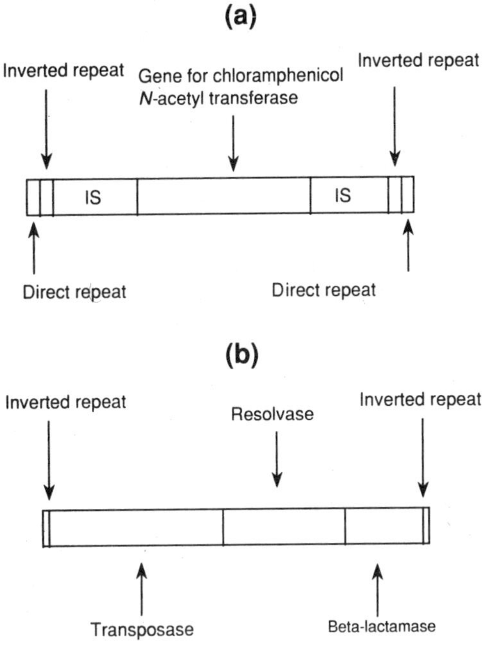

FIGURE 8.3 The structure of transposons: (a) Class 1 transposon Tn9, which includes the gene for bacterial resistance to chloramphenicol. IS, insertion sequence. (b) This class 2 complex transposon, Tn3, confers resistance to β-lactam antibiotics. The gene dimensions are not drawn to scale.

identified in both Gram-positive and Gram-negative bacteria.

In some transposons the drug-resistance genes are arranged within structures called integrons. These consist of an *int* gene that encodes for a site-specific recombination enzyme or integrase, an integron receptor site, *attI*, and one or more gene cassettes. Usually each gene cassette contains a single drug-resistance gene and a specific recombination site, called a 59-base-pair element, located downstream of the gene. The association of the integrase function with the specific recombination site confers mobility on gene cassettes and the ability of integrons to capture and integrate whole arrays of cassettes. Cassette excision is the reverse of integration and generates a circularized form of the cassette which may exist independently for extended periods. There are multiresistance integrons that confer various combinations of resistance to β-lactams, aminoglycosides, trimethoprim, chloramphenicol, antiseptics and disinfectants. More than 40 gene cassettes and three classes of integrons are known, and the reader is referred to reviews listed under 'Further reading' for detailed descriptions of this complex field. Although integrons are found in transposons, they also occur frequently as independent entities.

Mobilization of class 1 transposons along with their complement of drug-resistance genes occurs by nonreplicative transposition; that is, the transposon copy number is not increased during transposition. The transposon is excised from its original site and reinserts into a new site virtually anywhere within the bacterial genome, including chromosomal and plasmid sites. Class 2 transposons, on the other hand, are mobilized by a replicative process. The replicated copy of the transposon inserts into a new site. In both cases the transposase first introduces staggered cuts, nine base pairs apart, at the donor site in the transposon and at the intended recipient site. The recipient site, 4–12 base pairs in length, is then replicated to form noninverted, or direct repeats on either side of the inserted transposon. In contrast, as described earlier, gene cassette excision and capture in integrons is accomplished by site-specific recombination, although there are rare examples of cassettes that integrate into nonspecific sites.

To summarize, therefore, drug-resistance genes in bacteria are subject to two major modes of intragenic mobilization that promote a continual flux of resistance determinants around bacterial DNA:

1. Resistance genes associated with transposons, whether or not as cassettes, are mobilized along with the rest of the transposon and can be inserted essentially anywhere in the bacterial genome, either chromosome or plasmid.
2. Both transposon-associated and independent integrons containing resistance gene cassettes exchange and capture cassettes by site-specific recombination.

Conjugative transposons

The transposons described so far are by themselves unable to promote gene transfer by conjugation between bacterial cells, although they participate as passengers during R-plasmid transfers. However, there is another type of transposon, referred to as conjugative transposons. These are discrete DNA elements normally integrated into bacterial chromosomes which encode proteins that enable excision of the transposon from the chromosome and its transfer to recipient bacteria by intercellular conjugation. Conjugative transposons occur widely in Gram-positive bacteria and contribute to the spread of drug resistance among major pathogens such as *Streptococcus* spp. and *Enterococcus* spp. In Gram-negative bacteria, conjugative transposons were first identified in the genus *Bacteroides*, which accounts for 25–30% of the microbial flora of the human intestinal tract. Subsequently, conjugative transposons carrying drug-resistance genes have been found in other Gram-negative species, including *Salmonella, Vibrio* and *Proteus*.

The potential for conjugative transposons to spread drug resistance was highlighted when the first conjugative transposon to be discovered, Tn916, was found to carry resistance to tetracycline. Originally Tn916 was detected on the chromosome of a multiresistant isolate of *Enterococcus* (previously called *Streptococcus*) *faecalis*, and it was also observed to integrate readily into coresident plasmids and into many sites of the chromosomes of bacterial recipients of Tn916. A closely related conjugative transposon, Tn1545, found in *Streptococcus pneumoniae*, also mediates tetracycline resistance as well as resistance to erythromycin and kanamycin.

The mobilization of conjugative transposons from bacterial chromosomes involves the following steps, although details of the initial signals for mobilization are not fully defined:

1. Staggered cuts are introduced at each end of the transposon, leaving 6-nucleotide, single-stranded stretches of DNA, known as coupling sequences.
2. The noncomplementary coupling sequences are then ligated to generate covalently closed, double-stranded circular intermediates.
3. During the insertion stage, the coupling sequences form temporary non-base-pairing interactions with the target site, which can be in either a coresident plasmid or the chromosome of a recipient bacterium after conjugation. It is not yet clear how correct base pairing is subsequently established, but it may involve either replication through the insertion region or repair of a mismatch. The insertion process of conjugative transposons differs from that of 'true' transposons in that the recipient or target site is not replicated.

The intercellular conjugation process promoted by conjugative transposons is not well understood. Unlike the process mediated by R-plasmids in Gram-negative bacteria (see next section), surface pili do not appear to be involved. It is clear that only single-stranded copies of the transposons are transferred to recipient cells during conjugation.

The intercellular traffic of conjugative transposons is highly regulated. Many *Bacteroides* transposons carry the *tetQ* gene for tetracycline resistance, which is dependent upon ribosomal protection. Remarkably, tetracycline is a highly effective stimulant of conjugative transposon-mediated mating in these species. A suite of transposon genes activated by tetracycline promotes transposon mobilization and self-transfer as well as mobilization of coresident plasmids sharing the same donor cells. Tetracyclines can therefore stimulate the spread of the resistance genes throughout the bacterial population of the intestinal tract, including many other bacterial species.

Conjugative transposons comprise a highly variable group of mobilizable DNA elements, and the reader is directed to references under 'Further reading' for more detailed descriptions.

8.2.2 R-plasmids

Cellular conjugation mediated by R-plasmids is the major mechanism for the spread of drug resistance through Gram-negative bacterial populations. R-plasmids usually exist separately from the bacterial chromosome. They consist of two distinct but frequently linked entities:

1. The genes that initiate and control the conjugation process, and
2. A series of one or more linked genes, often found within transposon-integron com-

plexes, which confer resistance to antibacterial agents.

The conjugative region is closely related to the F-plasmid, which also confers on Gram-negative bacteria the ability to conjugate with cells lacking an F-plasmid. A complete R-plasmid resembles the F-prime plasmid (F′) in carrying genetic material additional to that which controls conjugation.

A great variety of R-plasmids have been described which carry various combinations of drug-resistance genes. Other phenotypic characteristics conferred by R-plasmids can be used in systems of classification. These include the ability (fi^+) or inability (fi^-) to repress the fertility properties of an F-plasmid coresident in the same cell; the type of sex pilus (see later discussion) that the R-plasmid determines; the inability of R-plasmids to coexist in a bacterium with certain other plasmids, which permits the division of R-plasmids into several incompatibility groups; and finally, the presence of genes in the R-plasmid that specify DNA restriction and modification enzymes. R-Plasmids are usually defined by a combination of these properties.

Molecular properties of R-plasmids

R-Plasmids can be isolated from host bacteria as circular DNA (Figure 8.4) in both closed and nicked forms, and both forms coexist in the cell. The closed circular structure is probably adopted by R-plasmids not engaged in replication. The contour lengths and thus molecular weights of isolated R-plasmids depend very much on the host bacterium and the culture conditions prevailing immediately before the isolation procedure. The R-plasmid may sometimes dissociate into its conjugative and resistance determinants. This is more common in some host species, e.g. *Proteus mirabilis* and *Salmonella typhimurium*, than in *Escherichia coli*, where dissociation is rare. Dissociation seems to depend on the activity of a simple transposon that may be inserted at the junction of the two regions. The molec-

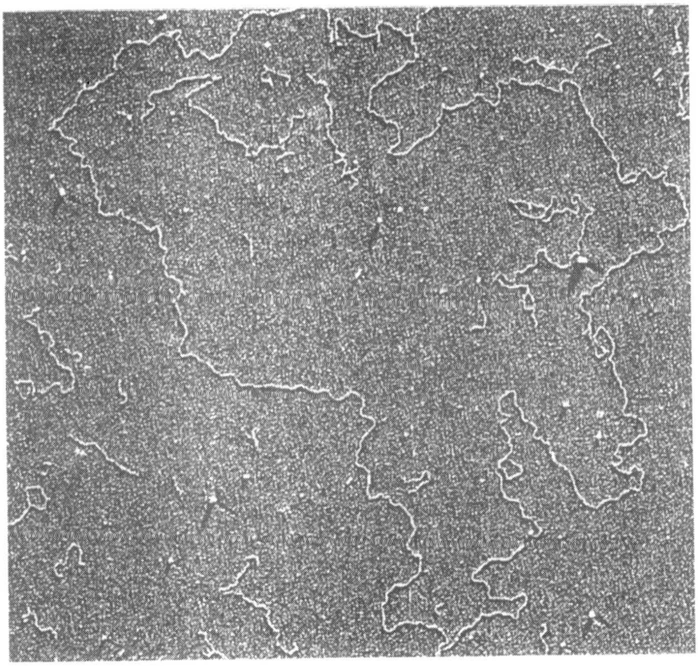

FIGURE 8.4 Electron micrograph of R-plasmid DNA isolated from *Proteus mirabilis* harbouring an R-plasmid with resistance markers to streptomycin, sulfonamides and chloramphenicol. The circular DNA has a total length of 28.5 μm. [This photograph is reproduced from *J. Bacteriol.* **97**, 383 (1969) by kind permission of Dr. Royston Clowes and the American Society for Microbiology.]

ular masses of between 50×10^6 and 60×10^6 kDa of the conjugative regions from R-plasmids are much greater than those of the drug-resistance genes. For example, the genes for chloramphenicol, streptomycin, spectinomycin and sulfonamide resistance have a combined molecular mass of only 12×10^6 kDa.

The copy number of R-plasmids harboured by individual bacteria is determined by the properties of the plasmid and its hosts as well as the culture conditions. As a general rule (to which there are exceptions), the larger R-plasmids are present in only a limited number of copies (between one and four) per chromosome in *Escherichia coli*, whereas in *Proteus mirabilis* the number is much more variable and even varies during the growth cycle. Conditions that give rise to an increased number of R-plasmid copies are sometimes associated with enhanced resistance. However, the level of cellular resistance does not always reflect the number of resistance gene copies. For example, although the number of R-plasmid copies is frequently greater in *Proteus mirabilis* than in *Escherichia coli*, the level of resistance to several drugs expressed in the former organism is usually lower than in *Escherichia coli*.

Cellular conjugation and R-plasmid transfer

Cells bearing an R-plasmid (R^+) are characterized by their ability to produce surface appendages known as sex pili. The sex pili of R^+ bacteria resemble those produced by F^+ organisms. When R^+ cells are mixed with sensitive R^- cells, mating pairs are immediately formed by surface interaction involving the sex pili. The transfer of a copy of the R-plasmid from the R^+ to the R^- cell begins, and the acquisition of the R-plasmid by the recipient cell converts it to a fertile, drug-resistant cell that can in turn conjugate with other R^- cells. In this way drug resistance spreads rapidly through the bacterial population. Uncovering the details of bacterial conjugation and the transfer of DNA has challenged investigators for many years. Although a wealth of information has emerged, several critical steps in the process remain to be defined. What may appear superficially to be a fairly simple phenomenon is in fact highly complex, and here we provide only an outline of the process.

The conjugal pair is brought into close surface contact by the attachment of the pilus of the donor cell to the recipient and its subsequent retraction by a process of 'reeling in'. The interaction between the cells triggers cleavage of a specific strand of the donor R-plasmid in the origin-of-transfer site (*oriT*) within a protein-DNA complex called the relaxosome, which contains the DNA strand-cleaving, or relaxase, enzyme. Only one strand, the T-strand, which is unwound following plasmid cleavage at the *oriT*, is transferred in a 5′ to 3′ direction from the donor to the recipient cell. Determining how the T-strand leaves the donor cell and penetrates the recipient has been a major research challenge. The extrusion of DNA from donor cells has some features in common with the secretion of toxins and virulence proteins referred to as the type IV secretory process. So-called coupling proteins are involved in transferring the exported proteins across the complex cell envelope of Gram-negative bacteria into the external environment. A recent suggestion is that in a typical plasmid such as R388, after generation of the T-strand of DNA, the relaxase protein, TrwC, serves as a pilot to guide the T-strand into the type IV secretory pore. A coupling protein, TrwB, then 'pumps' the DNA strand through the transporter-pore system, which perhaps involves ATP hydrolysis as an energy source. The details of the final transfer into the recepient cell remain shrouded in uncertainty, although the model suggests a possible role for the surface pili in breaching the permeability barriers of the recipient cells. It must be emphasized, however, that this proposal is speculative and is supported mainly by circumstantial evidence on the nature of the coupling and pilot proteins and the effects of loss-of-function mutations in these proteins on the DNA transfer process. Once inside the recipient cell, the ends of the transferred strand are ligated to produce covalently closed circular DNA. Finally, DNA replication, catalyzed by DNA polymerase III, generates double-stranded plasmid DNA from the single-stranded molecules in both donor and recipient cells.

Fortunately, the frequency of R-plasmid transfer is much lower than that of F transfer. Following the infection of an R^- cell with an R-plasmid, a repressor protein accumulates which eventually inhibits sex pilus formation. The ability to conjugate is therefore restricted to a short period immediately after acquisition of the R-plasmid. Sex pilus production in F^+ cells, in contrast, is not under repressor control and conjugal activity is therefore unrestricted. Mutant R-plasmids without the ability to restrict sex pilus formation exhibit a much higher frequency of R-plasmid transfer.

It is also worth noting that certain R-plasmids, and other self-mobilizing plasmids without drug-resistance genes, can promote the intercellular transfer of coresident plasmids that lack the genetic information for conjugation and transfer. Such mobilizable plasmids achieve transfer either by using the conjugal apparatus furnished by the self-mobilizing plasmids (*trans* mobilization) or by integration with these plasmids (*cis* mobilization). Cooperative interactions amongst plasmids add significantly to the genetic flexibility of bacteria and to their ability to spread drug resistance through microbial populations.

Clinical importance of R-plasmids

It is generally agreed that R-plasmids existed before the development of modern antibacterial drugs. Clearly though, the widespread use and abuse of these drugs led to a vast increase in drug resistance caused by R-plasmids. This has been especially noticeable in farm animals, which in many countries receive clinically valuable antibacterial drugs, or compounds chemically closely related to them, in their foodstuffs as growth enhancers. The animals act as a reservoir for Gram-negative bacteria, such as *Escherichia coli* and *Salmonella typhimurium*, harbouring R-plasmids potentially transferable to man. Fortunately, some countries have restricted the growth-enhancer application of clinically valuable antibiotics, although contravention of the regulations is not unknown.

The adverse contribution of R-plasmid-mediated drug resistance to human morbidity and mortality is undeniable. For example, the major requirement in the treatment of neonatal diarrhea caused by certain pathogenic strains of *Escherichia coli* (a potentially dangerous condition) is the prevention of fatal dehydration. Even so, elimination of the pathogenic organisms may also be important, but this is often difficult in the face of multiple resistance to commonly used antibacterial agents. In one notorious outbreak, the children were infected with a pathogenic strain of *Escherichia coli* resistant to β-lactams, streptomycin, neomycin, chloramphenicol and tetracyclines. The infection eventually responded to gentamicin, which was the only drug of those tested to which the pathogenic bacteria were sensitive. Another potentially alarming development has been the appearance of the typhoid organism, *Salmonella typhi*, carrying an R-plasmid with genes for resistance to chloramphenicol and cotrimoxazole, the drugs most commonly used to treat this disease.

Certain ecological factors probably limit the clinical threat posed by R-plasmids. In the environment of the gastrointestinal tract, the conjugal activity of R^+ bacteria may be less than that in the ideal culture conditions of the laboratory. The emergence of an R^+ population of bacteria during antibiotic therapy is more likely to result from selection of resistant cells than from extensive conjugal transfer of resistance. After cessation of antibiotic treatment, the numbers of R^+ bacteria in the feces fall, although usually not to zero. Low-level antibiotic contamination of the environment and/or a previously unsuspected persistence of drug-resistance and resistance-transfer genes in bacterial populations may contribute to this potentially serious situation.

8.2.3 Conjugative plasmids in Gram-positive bacteria

Although the existence of conjugative plasmids which carry drug-resistance genes in Gram-positive bacteria has been recognized for some time, their role in the dissemination of drug resistance is only now being more thoroughly investigated. As described earlier, the physical contact between Gram-negative bacteria necessary for the conjugal transfer of genes is largely attributable to the sex pili. A similar mechanism has not been identified in major Gram-positive pathogens. Major differences in the cell envelopes of Gram-positive and Gram-negative bacteria suggest that the modes of intercellular DNA transfer may also differ substantially. However, sequencing studies on several Gram-positive conjugative plasmids reveal homologies with proteins of the type IV secretion system involved in Gram negative R-plasmid transfer. Furthermore, the relaxosome of Gram-positive conjugative plasmids is similar to that of the R-plasmids.

8.2.4 Nonconjugal transfer of resistance genes

Transduction

During the two distinct processes of phage transduction, which occurs in both Gram-positive and Gram-negative bacteria, genetic information is transferred by phage particles from one bacterium to a related phage-susceptible cell.

Generalized transduction may occur during the lytic phases of both virulent and temperate phages. Fragments of degraded host chromosomal and plasmid DNA, which may carry drug-resistance determinants, can become packaged into newly generated phage particles, leaving behind some or all of the phage DNA. Lytic release of the phages enables them to inject both phage and donor host DNA into other bacteria, some of which is integrated into the recipient genome, although between 70 and 90% of the transferred DNA is not integrated in this way. Nevertheless, nonintegrated DNA may also survive in the recipient and replicate as a plasmid. In abortive transduction, none of the transferred DNA is integrated into the recipient genome, but again, the nonintegrated DNA survives and replicates as a plasmid. The drug-resistant phenotype in recipient bacteria is maintained in both types of transduction.

The process of specialized transduction depends on an error in the lysogenic cycle. Excision of the phage DNA from the host genome during induction of the lytic phase is insufficiently precise and carries some of the bacterial DNA along with phage DNA. The resulting phage genome contains up to 10% of the bacterial DNA next to the phage integration site in the bacterial genome. Clearly this process has the potential for generating infectious phage particles that carry bacterial genes for drug resistance. Although recombinant or defective phages arising from specialized transduction are able to inject their DNA into new hosts, they cannot reproduce independently, nor are they lysogenic. Specialized transduction has been most thoroughly studied with lambda phage, and the reader is referred to a relevant text on bacterial genetics for a detailed account of the mechanisms involved in lambda phage transduction. In general terms, relatively little of the injected recombinant phage DNA is integrated into the bacterial genome unless the phage population contains normal phages as well as the defective ones. The normal phages insert into the bacterial genome at a specific *att* site that resembles the phage *att* site. The insertion process generates two hybrid bacterial-phage *att* sites where the defective recombinant phage DNA can insert. The presence of the normal phage renders the bacteria susceptible to the induction of phage-mediated lysis. The resulting lysate, containing roughly equal amounts of defective recombinant phages and normal phages, is highly efficient in transduction.

Although transduction of drug-resistance determinants is readily demonstrated under laboratory conditions, its contribution to the spread of drug resistance in natural and clinical settings is difficult to quantify.

Transformation

Under certain conditions most genera of bacteria can absorb, integrate and express fragments of 'naked' DNA containing intact genetic information, including that for drug resistance. The phenomenon of transformation of bacteria by DNA is more complex than it may appear at first glance. It has been most thoroughly investigated in the species in which it was first discovered more than 60 years ago, *Streptococcus pneumoniae*. Only bacteria in a state of competence are able to absorb and integrate exogenous DNA into their own genome. *Streptococcus pneumoniae* becomes competent during exponential growth when the population density exceeds 10^7–10^8 cells ml^{-1}. Under these conditions the bacteria secrete a competence factor that stimulates the synthesis of up to 10 other proteins essential for transformation. Competent cells bind double-stranded DNA, provided that its molecular mass is at least 500 kDa. One strand of the DNA is hydrolyzed by an exonuclease associated with the cell envelope and the remaining strand enters the cell while bound to competence-specific proteins. Integration into a homologous region of the recipient genome probably occurs by nonreciprocal general recombination.

While competent *Streptococcus pneumoniae* can take up DNA from a range of bacterial species, the Gram-negative opportunist pathogen *Haemophilus influenzae* is more fastidious and only accepts DNA from closely related species. Furthermore, *Haemo-*

philus influenzae does not produce a competence factor but absorbs double-stranded DNA encapsulated in membrane vesicles. Although transformation is a widespread phenomenon, it is not surprising to find important differences in the details of the actual mechanism among the bacterial species. The complexity and diversity of the transformation system indicates its evolutionary importance in the exchange of genetic information in the bacterial world. The frequency of transformation of genetic markers can be as high as 10^{-3} under laboratory conditions when artificially high levels of DNA are added, i.e. one cell in every thousand takes up and integrates a particular gene. Transformation is probably therefore a significant contributor to the spread of drug-resistance genes. A specific example of the relevance of transformation to drug resistance is illustrated by the existence of the mosaic genes for PBPs with diminished affinity for β-lactam antibiotics, for sulfonamide-resistant dihydropteroate synthase and for the *tetM* form of tetracycline resistance referred to previously. Unfortunately, the extent to which transformation occurs in relevant environments, such as hospital wards and the intestinal tract, cannot be quantified with any certainty.

8.3 Global regulators of drug resistance in Gram-negative bacteria

We have seen how the capture of several determinants for resistance to individual drugs by mobile and transferable genetic elements can result in bacteria acquiring the multidrug-resistant phenotype. However, Gram-negative bacteria have yet another means of achieving a similar end. Genes called regulons exert transcriptional control over several chromosomal genes that confer resistance to many antibiotics by restricting access to their molecular targets. The *marA* locus in *Escherichia coli* contains an operon, *marRAB*, whose expression is inducible by at least two antibiotics, tetracycline and chloramphenicol, and by uncouplers of oxidative phosphorylation. Resistance to these and to many other drugs is increased by the induction process. The MarA protein encoded in the operon belongs to a family of transcriptional regulators and controls the expression of numerous other genes, probably in concert with MarR, a regulator protein, and the MarB protein, whose function is currently unknown. The *mar* operon, when induced, upregulates the multidrug efflux pump AcrB and its linked outer membrane component, TolC (Chapter 7). The *mar* operon also increases the expression of *micF*, an RNA molecule which slows the expression of the OmpF porin channel protein. Enhanced drug efflux combined with diminished outer membrane permeability caused by a reduction in the expression of OmpF underlies the increase in resistance to a range of structurally unrelated drugs that is mediated by the *marA* locus.

Another global regulator of drug resistance, *ramA*, has been found in the Gram-negative pathogen *Klebsiella pneumoniae*. The *ramA* gene encodes a transcriptional activator protein, RamA, that is distantly related to the MarA protein of *Escherichia coli*. Like MarA, RamA confers resistance to a wide range of structurally unrelated drugs by upregulating the expression of several genes. RamA-mediated resistance also appears to depend upon a combination of drug efflux and a reduction in the level of the OmpF protein in the outer membrane.

Genetic loci resembling *marA* and *ramA* are widespread among Gram-negative bacteria, and the regulation of multidrug resistance by these operons is likely to be a significant contributor to the overall problem of drug resistance in Gram-negative bacteria.

8.4 Genetic basis of resistance to antifungal drugs

The yeast pathogen *Candida albicans* exists exclusively as a diploid organism, i.e. it has two copies of every gene. Allelic differences between the two gene copies are commonly encountered among clinical isolates of *Candida albicans*. Resistance to the azole antifungal drugs is often caused by mutations in the target enzyme 14α-demethylase (Chapter 3) encoded by *erg11*. For example, there is a point mutation (R467K) which replaces lysine with arginine at position 467. Analysis of azole-resistant clinical isolates of *Candida albicans* showed that resistance was high when both gene copies carried the R467K mutation compared with heterozygotes in which only one allele was mutated. Resistance to the nucleoside analogue, 5-fluorocytosine in *Candida albicans* is also modulated by allelic differences. As described in Chapter 4, 5-FC is first converted in the cells to 5-fluorouracil by the enzyme cytidine deaminase. A mutation in one allele for this enzyme results in partial resistance to 5-FC because the unaffected allele continues to express the wild-type enzyme. The generation of cells homozygous for the disabled gene via mitotic recombination causes high-level resistance to 5-FC. In contrast, the haploid yeast *Candida neoformans* can acquire high-level resistance to 5-FC in a single step because there is only one copy of the gene for cytidine deaminase. This situation is reflected in the clinic, where high-level resistance to 5-FC is more commonly detected in *Candida neoformans* than in *Candida albicans*.

Transposable genes and plasmids occur widely in fungi, especially in yeasts. However, at present there is no evidence that these elements contribute to the spread of drug resistance among fungal pathogens such as *Candida* spp., and there is no phenomenon in these organisms comparable with that of mobilizable and transmissable drug-resistance genes in bacteria.

8.5 Genetic basis of drug resistance to antimalarial drugs

As described in Chapter 9, the biochemical basis of resistance to antimalarial drugs can be due either to reduced drug uptake by the parasite, as in the case of chloroquine and possibly other quinoline drugs, or to the loss of drug sensitivity in target enzymes, including dihydrofolate reductase and dihydropteroate synthase. Spontaneous mutations resulting in the selection of drug-resistant parasites in response to sustained drug challenge appear to be a common pattern throughout the malarious regions of the world. Sexual reproduction and genetic recombination in the malarial parasite provide additional opportunities for the spread of drug resistance.

Further reading

Balis, M. M. *et al.* (2002). Mechanisms of fungal resistance. *Drugs* **62**, 1025.

Chopra, I. and Roberts, M. (2001). Tetracycline antibiotics: mode of action, applications, molecular biology and epidemiology of bacterial resistance. *Microbiol. Molec. Biol. Rev.* **65**, 232.

Courvalin, P. (1994). Transfer of antibiotic resistant genes between Gram-positive and Gram-negative bacteria. *Antimicrob. Agents Chemother.* **38**, 1447.

Errington, J. *et al.* (2001). DNA transport in bacteria. *Nat. Rev. Mol. Cell. Biol.* **2**, 538.

George, A. M., Hall, R. M. and Stokes, H. W. (1995). Multidrug resistance in *Klebsiella pneumoniae*: a novel gene *ramA* confers a multidrug resistance phenotype in *Escherichia coli*. *Microbiol.* **141**, 1909.

Gomis-Ruth, F. X. *et al.* (2002). Structure and role of coupling proteins in conjugal DNA transfer. *Res. Microbiol.* **153**, 199.

Grohmann, E. *et al.* (2003). Conjugative plasmid transfer in Gram-positive bacteria. *Microbiol. Mol. Biol. Rev.* **67**, 277.

Henriques-Normark, B. and Normark, S. (2002). Evolution and spread of antibiotic resistance. *J. Int. Med.* **252**, 91.

Le Bras, J. L. and Durand, R. (2003). The mechanisms of resistance to antimalarial drugs in *Plasmodium falciparum*. *Fundam. Clin. Pharmacol.* **17**, 147.

Llosa, M. *et al.* (2002). Bacterial conjugation: a two-step mechanism for DNA transport. *Molec. Microbiol.* **45**, 1.

Lodish, H., Berk, A., Zipursky, S. L., Matsudaira, P., Baltimore, D. and Darnell, J. (2000). *Molecular Cell Biology*. 4th edn. W. H. Freeman & Co., New York.

Pembroke, J. T. *et al.* (2002). The role of conjugative transposons in the Enterobacteriaceae. *Cell. Mol. Life Sci.* **59**, 2055.

Pillay, D. (2001). The emergence and epidemiology of resistance in the nucleoside-experienced HIV-infected population. *Antivir. Ther.* **6** Suppl. 3, 15.

Prescott, L. M., Harley, J. P. and Klein, D. A. (2004). *Microbiology*, 5th edn, William C. Brown, Dubuque, IA.

Recchia, G. D. and Hall, R. M. (1995). Gene cassettes: a new class of mobile element. *Microbiol.* **141**, 3015.

Chapter nine

Biochemical mechanisms of resistance to antimicrobial drugs

Although the individual modes of resistance to antimicrobial drugs are diverse, they can be grouped into a set of general mechanisms that account for most types of resistance encountered in medical practice.

1. Conversion of the active drug to an inactive derivative by enzyme(s) synthesized by the resistant cells;
2. Loss or downregulation of an enzymic mechanism required to convert an inactive drug precursor to an active antimicrobial agent;
3. Loss of sensitivity of the drug target site as a result of: (a) modification of the target site by enzyme activity in the resistant cells, (b) mutation(s) in the microbial chromosome affecting the target, (c) horizontal acquisition of genetic information encoding a drug-resistant form of the target enzyme, overproduction of the drug-sensitive enzyme, or protein(s) that protect the target site from inhibition;
4. Removal of the drug from the cell by drug efflux pumps.
5. Reduction in cellular permeability to drugs caused by changes in the cell envelope.

The actual level of cellular resistance observed may be due to a combination of factors. In Gram-negative bacteria, for example, resistance often results from the low permeability of the outer membrane combined with drug efflux and other mechanisms. In addition there are unique modes of resistance not included in this broad classification but which are nevertheless of major medical importance, e.g. vancomycin resistance.

The rest of this chapter is devoted to examples of the biochemical processes involved in resistance to clinically important drugs used in the treatment of infections caused by bacteria, fungi, viruses and protozoa.

9.1 Enzymic inactivation of drugs

9.1.1 β-Lactams

The destruction of penicillins, cephalosporins and carbapenems by bacteria that produce β-lactamases is one of the most widespread and serious forms of microbial resistance. The general inactivation reactions are shown in Figure 9.1. The β-lactam bonds of penicillins and cephalosporins are cleaved to yield the

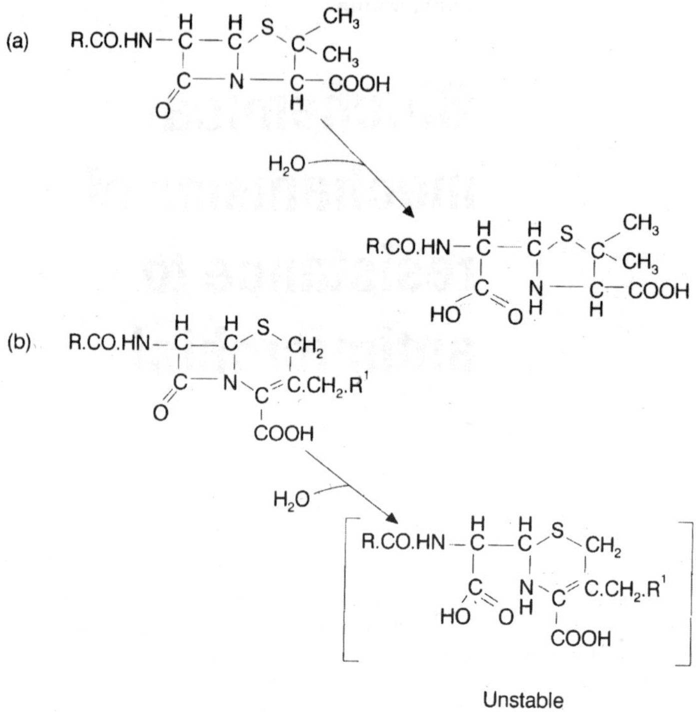

FIGURE 9.1 Inactivation of (a) penicillins and (b) cephalosporins by β-lactamase. Whereas penicilloic acid is relatively stable, the corresponding cephalosporin product is highly unstable and decomposes spontaneously to a complex mixture. R and R[1] represent a wide variety of side chains that may substantially affect the efficiency of β-lactamase attack.

biologically inactive derivatives penicilloic acid and cephalosporanoic acid, respectively. Penicilloic acid is a stable end product, but cephalosporanoic acid spontaneously degrades to several other compounds. As we shall see, the nature of the R-substituent in the amide side chain of β-lactams is important in determining the susceptibility of compounds to β-lactamase attack. The number of β-lactamases produced by different bacteria is astonishing—more than 340 such enzymes have been identified so far, with a wide range of substrate preferences for penicillins, cephalosporins and carbapenems. Clinical isolates of bacteria commonly express several different β-lactamases, thus providing a formidable array of defences against many different antibiotics.

Classification of β-lactamases

The numbers and diversity of β-lactamases are remarkable; as mentioned previously, at least 340 bacterial β-lactamases have been described. The classification of this wealth of enzymes is a major challenge and there are two current schemes: the Ambler classification based on similarities in amino acid sequences, and the Bush-Jacoby-Medeiros system based on substrate and inhibitor profiling. In the Ambler scheme there are four classes:

Class A: penicillinases and cephalosporinases usually found on plasmids or transposons.
Class B: metallo β-lactamases.
Class C: chromosomal cephalosporinases
Class D: oxacillinases

The Bush-Jacoby-Medeiros scheme also has four classes or groups:

Group 1: chromosomally encoded cephalosporinases that are poorly inhibited by clavulanic acid.

Group 2: penicillinases, cephalosporinases and broad-specificity enzymes, including carbapenemases, both chromosomally and plasmid encoded that are inhibited by clavulanic acid and other active-site-directed inhibitors of β-lactamases

Group 3: metallo-β-lactamases unaffected by all the conventional β-lactamase inhibitors

Group 4: a limited number of uncharacterized penicillinases that are not inhibited by clavulanic acid

Group 2 includes an extended and varied set of enzymes that are further subdivided into eight subclasses according to their substrate and inhibitor profiles. Assuming that novel forms of β-lactamases continue to be discovered, further revision of the existing schemes of classification may eventually be needed.

Gram-positive β-lactamases

The most important β-lactamase in Gram-positive bacteria is that produced by *Staphylococcus aureus*, which was responsible for the rise in the resistance of this pathogen to penicillin first observed in the late 1940s and 1950s. In many hospitals today more than 90% of the *Staphylococcus aureus* isolates are resistant to the simpler penicillins because of β-lactamase. The β-lactamase of *Staphylococcus aureus* is an inducible enzyme. Enzyme production is very low in the absence of penicillin or cephalosporin. The addition of minute quantities of antibiotic (as little as 0.0024 μg ml^{-1} of medium) increases enzyme production enormously, and the β-lactamase may account for more than 3% of the total protein synthesized by the bacterium. The enzyme, which preferentially attacks penicillins, is released from the bacterial cell and inactivates antibiotic in the surrounding medium. The resulting dilution of β-lactamase is the basis of the well-known 'inoculum effect'. A small inoculum of *Staphylococcus aureus* cells may not destroy all the antibiotic in the medium, but the much greater quantity of enzyme produced by a heavy inoculum of cells is able to overcome the challenge. Staphylococcal resistance to penicillin is therefore dependent upon the size of the inoculum.

The regulatory genes which control the expression of the β-lactamase protein, BlaP (BlaZ in *Staphylococcus* aureus), are *blaI*, *blaR1* and *blaR2*. BlaI, the protein encoded by *blaI*, is a repressor that binds to operator sites between *blaI* and *blaP/Z* and prevents the expression of β-lactamase. BlaR1 is a large membrane-spanning protein of 601 amino acid residues with a 261-amino acid extracellular domain that interacts with β-lactams in the external environment. The recognition site is located mainly in the region around serine-402 (serine-389 in *Staphylococcus aureus*), which has close homology with the domains flanking the active-site serine residue of β-lactamase. Recent work with *Staphyloccocus* aureus shows that β-lactam antibiotics *O*-acylate the hydroxyl group of serine-389, which is activated by the nearby presence of the carboxylated side chain of lysine-392. A similar type of serine activation is also observed at the active center of β-lactamase. The acylation of BlaR1 by β-lactams initiates a conformational change in the extracellular domain of BlaR1 which transmits to a cytoplasmic region of the protein via the transmembrane domains. The cytoplasmic region of BlaR1 has a zinc-dependent protease domain which is activated by the conformational change and then cleaves the repressor protein BlaI. The repressor function of BlaI is eliminated by the proteolytic cleavage, thus permitting expression of the structural *blaP* gene and production of β-lactamase. The function of the BlaR2 protein at present is not known, but it is nevertheless essential for the regulation of β-lactamase synthesis. In *Staphylococcus aureus*, the genes encoding the synthesis and regulation of β-lactamase are frequently found on plasmids.

Gram-negative β-lactamases

The complex outer envelope of Gram-negative cells makes them intrinsically less sensitive to many β-lactams. However, after the introduction of penicillin derivatives such as ampicillin with good activity against Gram-negative bacteria, β-lactamase-mediated resistance amongst these pathogens soon began to emerge.

Many β-lactamases in Gram-negative bacteria are not inducible and are expressed constitutively. In *Escherichia coli*, β-lactamase is encoded by the chromosomal *ampC* gene, which is expressed constitutively at low levels. However, *ampC* mutations which confer overproduction of β-lactamase are observed at the low frequency of approximately 10^{-9}. The genetic changes responsible are often point mutations or insertions that increase the transcription of the *ampC* gene. There are also attenuator mutations that allow increased transcriptional readthrough into *ampC* and examples of amplification of the *ampC* gene. In some Gram-negative species, including *Enterobacter cloacae, Citrobacter freundii* and *Pseudomonas aeruginosa,* the enzyme is inducible, although the mechanism of induction differs from that in Gram-positive bacteria. Whereas β-lactamase in these species is normally expressed at low constitutive levels, mutations causing high levels of constitutive enzyme synthesis occur at much higher frequencies than in *Escherichia coli*: 10^{-6}. The expression of the *ampC* gene in *Enterobacter cloacae* and *Citrobacter freundii* is controlled by an adjacent *ampR* gene. Unlike the system in Gram-positive bacteria, β-lactams are not directly involved in the induction process in Gram-negative species. Under normal growth conditions, the *ampC* gene-activating protein, AmpR, interacts with a cytosolic intermediate in the biosynthesis of peptidoglycan, UDP-muramylpentapeptide (Chapter 2), which blocks the activating function of AmpR.

Inhibition of peptidoglycan cross-linking by β-lactams causes an increase in the normal turnover of peptidoglycan in Gram-negative bacteria. The turnover process involves the transport of peptidoglycan fragments generated in the cell wall across the cytoplasmic membrane into the cytoplasm. The transport process is effected by a transmembrane protein encoded by the *ampG* gene that is also essential for the induction of β-lactamase. The peptidoglycan fragments generated by the action of β-lactams displace UDP-muramylpentapeptide from its interaction with AmpR, thus allowing AmpR to exert its activating function and enhance the expression of *ampC*. Another gene, *ampD,* encodes an amidase, AmpD, that hydrolyzes the activating peptidoglycan fragment shown in Figure 9.2. The tripeptide released in this reaction is normally directly reutilized in peptidoglycan synthesis. Mutations that inactivate *ampD* result in high-level constitutive β-lactamase biosynthesis, owing to accumulation of the activating fragment. The negative regulatory activity of the amidase specified by wild-type *ampD* is presumably overwhelmed by the sudden influx of peptidoglycan fragments when the cell is under attack by β-lactams.

Catalytic mechanisms of β-lactamases

Although there is considerable diversity in amino acid sequences among the many known β-lactamases, the majority are acyl serine transferases. A smaller group includes zinc-dependent enzymes. The carbonyl carbon of the β-lactam amide bond transfers to a serine residue at the active center of the acyl serine transferase to form a serine ester-linked β-lactamoyl enzyme complex (Figure 9.3). This reaction is facilitated by activation of the serine γ-hydroxyl group by the carboxylated side chain of a nearby lysine residue in a manner comparable to the interaction of β-lactams with the BlaR protein described previously. The proton of the serine γ-hydroxyl group is abstracted, enabling the γ-oxygen atom to attack the carbonyl group of the β-lactam molecule. The abstracted proton is transferred to the adjacent nitrogen atom. In the next stage a proton is abstracted from a water molecule at the active center and the activated hydroxyl group attacks the serine-lactamoyl ester bond. The ensuing hydrolysis releases free enzyme and a biologically inactive derivative, either penicilloic or cephalosporanoic acid. This reaction is comparable to that between β-lactams and the penicillin-binding proteins involved in peptidoglycan biosynthesis (Chapter 2). In the latter case, the β-lactamoyl-PBP complexes, unlike the corresponding complexes in β-lactamases, are relatively resistant to attack by water molecules, resulting in long-lasting inactivation of the PBPs. The analogous interactions of the active-site serine residues of β-lactamases and PBPs with β-lactams, and similarities in the active-site sequences and the secondary structures of both proteins, suggest that β-lactamases and PBPs may have evolved from the same ancestral protein.

Several species of pathogenic bacteria, including *Klebsiella pneumonia, Pseudomonas* spp., *Bacteroides* spp., *Serratia marcescens* and *Acinetobacter* spp. produce chromosomally mediated β-lactamases with a zinc ion at the active center. Until relatively recently these metallo-β-lactamases were considered to be little more than interesting biochemical curiosities. However, their ability to degrade penicillins, cephalosporins and carbapenems, together with their lack of susceptibility to inhibitors of the acyl serine transferase β-lactamases, has highlighted the threat of

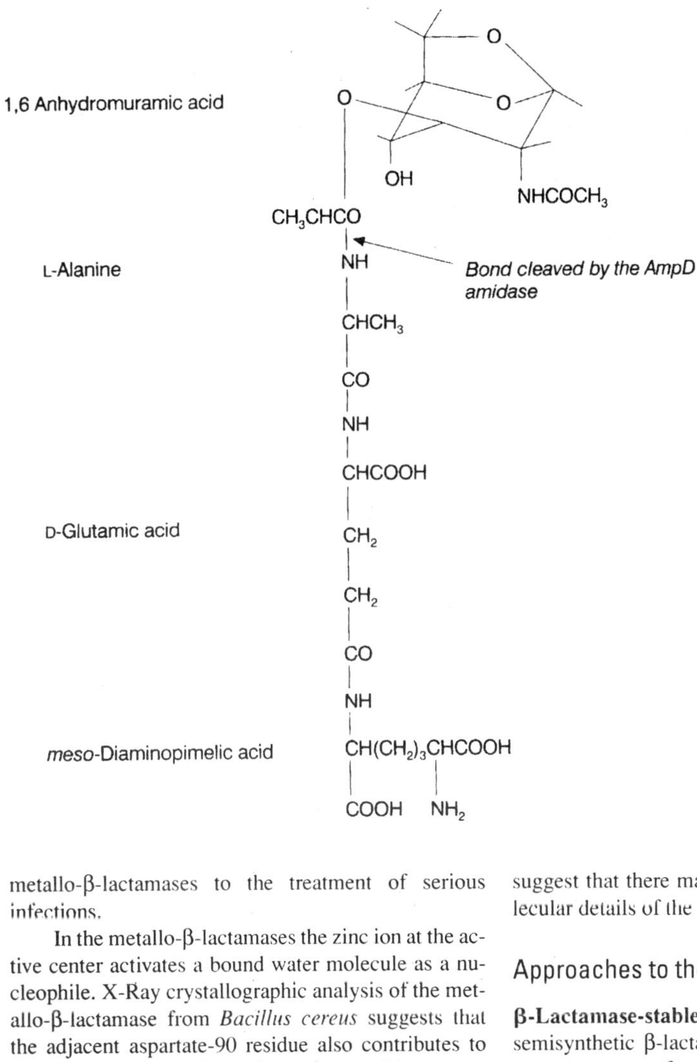

FIGURE 9.2 Structure of the molecule arising from the turnover of peptidoglycan that participates in the induction of the *ampC* β-lactamase of Gram-negative bacteria. Cleavage of the compound by the AmpD amidase inactivates the inducing activity. [Adapted with permission from C. Jacobs *et al. Molec. Microbiol.* **15**, 55 (1995). Published by Blackwell Science, London.]

metallo-β-lactamases to the treatment of serious infections.

In the metallo-β-lactamases the zinc ion at the active center activates a bound water molecule as a nucleophile. X-Ray crystallographic analysis of the metallo-β-lactamase from *Bacillus cereus* suggests that the adjacent aspartate-90 residue also contributes to the activation process by acting as a general base to remove a proton from the water molecule. The abstracted proton is donated to the nitrogen atom of the β lactam bond and cleavage of the bond ensues. Although most, if not all, of the recognized metallo-β-lactamases are believed to involve a zinc ion in the catalytic process, significant differences in the amino acid sequences and tertiary structures among the enzymes suggest that there may also be differences in the molecular details of the catalytic process.

Approaches to the β-lactamase problem

β-Lactamase-stable compounds. The advent of the semisynthetic β-lactams during the 1950s offered an apparent escape from the problem of staphylococcal resistance caused by β-lactamase. Compounds such as methicillin and cloxacillin (Chapter 2), with bulky substituents in the penicillin side chain, were found to be poor substrates for β-lactamase. The affinity of methicillin for staphylococcal β-lactamase is much lower than that of benzylpenicillin, and the maximum rate of hydrolysis of methicillin by this enzyme is only one-

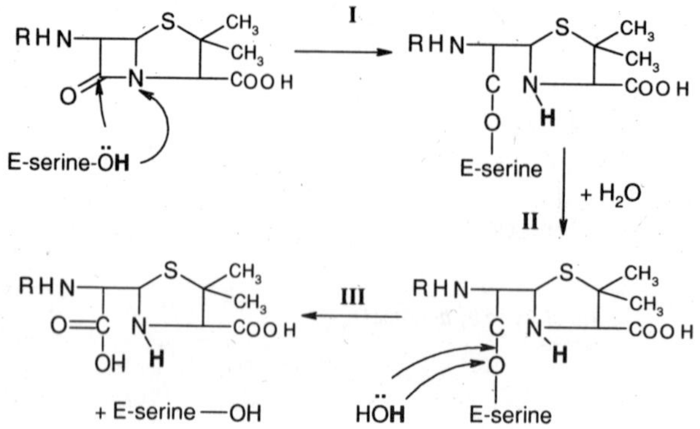

FIGURE 9.3 The essential reactions at the active centre of the serine β-lactamases. I: The proton of the γ-OH of the active-centre serine is abstracted and the resulting activated γ-oxygen atom attacks the β-lactam carbonyl group to form an ester link. The proton is then back-donated to the adjacent nitrogen atom. II: Abstraction of a proton from a water molecule at the active centre results in an attack of the activated water OH group on the serine-lactamoyl ester bond to release the degraded β-lactam and regenerated enzyme (III). E represents the rest of the enzyme molecule.

thirtieth of that of benzylpenicillin. Until relatively recently, methicillin was effective against infections caused by β-lactamase-producing staphylococci, even though its intrinsic antibacterial activity is substantially lower than that of benzylpenicillin. Although it is only slowly degraded by Gram-negative β-lactamases, methicillin is ineffective against Gram-negative infections because its physical characteristics limit its ability to penetrate the outer membrane. To combat the menace of Gram-negative β-lactamases, therefore, compounds were needed that both resisted β-lactamase attack and penetrated effectively to the PBPs in the cytoplasmic membrane.

An extensive range of novel β-lactam derivatives has been developed, with good activity against *Escherichia coli* strains producing the most commonly encountered β-lactamase of Gram-negative bacteria, TEM-1. This enzyme, whose name is derived from that of a patient treated for a β-lactam-resistant infection, is encoded on an R-plasmid that transmits readily to other enterobacteria. The examples in Table 9.1 show that *Escherichia coli* cells producing TEM-1 are effectively inhibited by β-lactamase-resistant compounds.

Extended spectrum β-lactamases. Unfortunately, following the introduction of novel β-lactams, the plasmid-borne enzymes in *Escherichia coli,* TEM-1 and TEM-2, underwent mutations near the active center that markedly increased their ability to hydrolyze several of these valuable agents. A third β-lactamase, SHV-1, which originated in *Klebsiella* spp. and conferred resistance to ampicillin, eventually transferred to *Escherichia coli* and evolved to hydrolyze novel β-lactams. There are now more than 90 of these extended-spectrum β-lactamases (ESBLs), with more than 70 in the TEM family and 20 or more in the SHV group. The ESBLs include both serine-active site and metallo-β-lactamases. Many structural variants of β-lactams are hydrolyzed, although the individual enzymes exhibit significant substrate specificities. It is interesting that ESBLs are frequently more sensitive to β-lactamase inhibitors, although inhibitor-resistant variants have appeared in both the United States and Europe. Two examples of ESBLs—TEM-12 and TEM-26—are listed in Table 9.1. TEM-26 has two critical amino acid replacements: serine-for-arginine at position 164 and lysine-for-glutamate at position 104. These changes dramatically enhance the hydrolytic efficiency of the enzyme against drugs of major importance such as ceftazidime and cefuroxime, and bacteria equipped with this enzyme are markedly more resistant to these drugs (Table 9.1). The carbapenems, in which the sulfur atom of the β-lactam fused ring system is replaced by a carbon atom, e.g. thienamycin and meropenem (Chapter 2), generally have good stability to the serine-active-site β-lactamases and are highly active against bacteria producing the various forms of TEM (see imipenem in Table 9.1). However, it is now apparent that carbapenems are hydrolyzed by some ESBLs, including both metallo-β-lactamases, and by some unusual serine-active-site enzymes.

Some bacteria, such as *Enterobacter cloacae,* have evolved a different strategy to combat β-lacta-

TABLE 9.1 The effects of TEM β-lactamase and two mutant variants on the hydrolysis of and bacterial sensitivity to β-lactams

Enzyme	*Relative rates of hydrolysis (benzylpenicillin = 100)*				*MIC values for* Escherichia. coli *(μg/ml)*			
	AMP	CTAX	CTAZ	IMP	AMP	CTAX	CTAZ	IPM
R^-	–	–	–	–	4	0.125	0.25	0.25
TEM-1	110	0.07	0.01	< 0.01	> 256	0.125	0.125	0.25
TEM-12	14	2.4	3.8	> 1	> 256	0.5	64	0.25
TEM-26		7.5	170		> 256	64	128	0.25

Notes: AMP, ampicillin; CTAX, cefotaxime; CTAZ, ceftazidime; IPM, imipenem. Although no value for the rate of hydrolysis of AMP by TEM-26 is presented, the high minimal inhibitory concentration (MIC) against *Escherichia coli* expressing this enzyme indicates rapid destruction of the antibiotic. R^- indicates bacteria without a resistance plasmid.

mase-resistant antibiotics. As discussed previously, the chromosomal *ampC* gene encoding β-lactamase in this organism is normally an inducible gene that is indirectly regulated via the action of β-lactams on peptidoglycan metabolism. In the modified strategy, the challenge of β-lactamase-stable drugs is countered by mutations in the regulatory cascade that markedly increase the production of enzyme. Large quantities of a β-lactamase with weak activity degrade enough antibiotic to allow the bacteria to survive and proliferate.

Inhibitors of β-lactamase. Some of the early β-lactamase-stable β-lactams, like cloxacillin, have a degree of inhibitory activity against β-lactamases. However, it was the discovery of a naturally occurring inhibitor of these enzymes, clavulanic acid (Chapter 2), that opened the way for effective synergism with β-lactamase-susceptible drugs such as ampicillin and amoxycillin. Although clavulanic acid has little intrinsic antibiotic activity, it is a remarkably effective inhibitor of many β-lactamases of Gram-positive and Gram-negative bacteria. Clavulanic acid and other clinically useful compounds, including sulbactam and tazobactam, react irreversibly with the active-site serine of β-lactamase to form stable, enzymically inactive complexes. However, bacterial evolution is again proving equal to the challenge of inhibitors of β-lactamase. Mutant forms of the enzyme have emerged which are highly resistant to inhibition. Numerous inhibitor-resistant forms of the TEM β-lactamases have been identified in clinical isolates of *Escherichia coli, Klebsiella pneumoniae* and *Proteus mirabilis.* Replacement of methionine at position 69 by the aliphatic amino acids isoleucine, leucine or valine, and of arginine-244 by serine or cysteine, cause marked increases in resistance to inhibitors. Other bacteria resistant to combinations of β-lactamase inhibitors with susceptible β-lactams are characterized by overproduction of β-lactamase, which overwhelms the protective capacity of the inhibitors.

The clinical threat posed by the metallo-β-lactamases continues to challenge the ingenuity of medicinal chemists to devise inhibitors of these enzymes. Unfortunately, at present there are no clinically useful inhibitors of metallo-β-lactamases, although there are some promising developments at the laboratory stage.

The three-dimensional structures of both serine-active-site and metallo-β-lactamases have been solved by X-ray crystallography. Two examples are illustrated in Figures 9.4(a) and 9.4(b). It is hoped that the information provided by the X-ray data together with synthetic organic chemistry will result in a continuing flow of novel agents to contain the threat of drug-resistant, β-lactamase-producing bacteria.

9.1.2 Chloramphenicol

The potential toxicity of chloramphenicol limits its use mainly to the treatment of life-threatening infections, such as typhoid and meningitis, where bacterial resistance to the drug may have serious consequences. Resistance to chloramphenicol is primarily due to the enzyme chloramphenicol acetyl transferase (CAT) which is widespread amongst most genera of Gram-positive

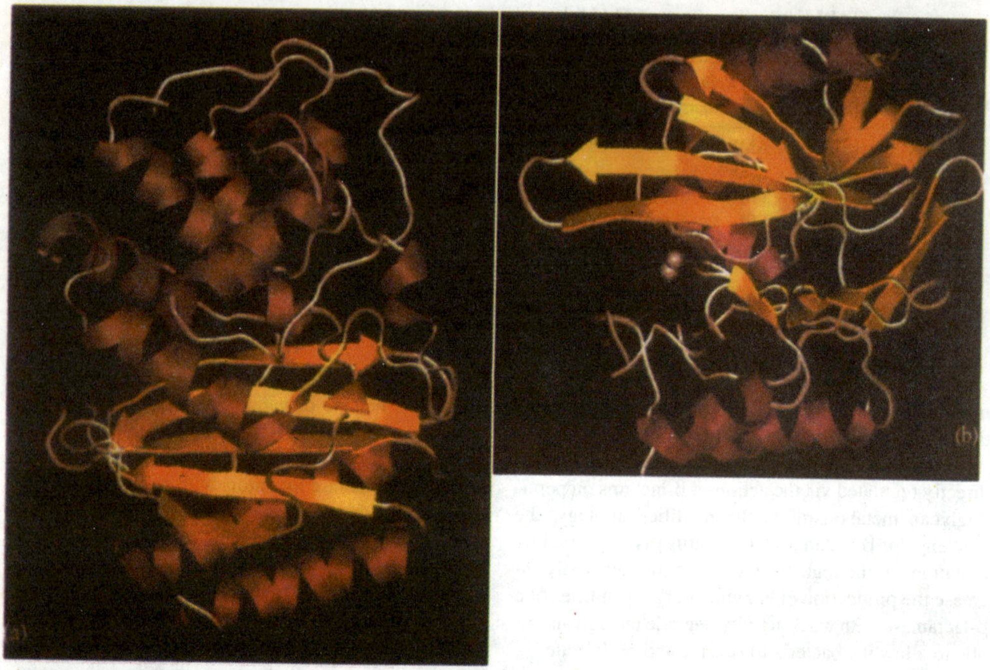

FIGURE 9.4 Three-dimensional structures, as revealed by X-ray crystallographic analysis. (a) The TEM-1 serine-active-center β-lactamase from *Escherichia coli.* [Reproduced with permission from C. Jelsch *et al. Proteins: Structure, Function and Genetics* **16**, 364 (1993). Published by Wiley-Liss, Inc., a subsidiary of John Wiley & Sons.] (b) The zinc β-lactamase from *Bacillus cereus.* [Reproduced with permission from A. Carfi *et al. EMBO J.* **14**, 4914 (1995). Published by Oxford University Press.] In both diagrams the spiral ribbons represent α-helices; the arrowed ribbons, β-pleated sheets; and the strings, looped regions of the proteins. The zinc atoms at the active centre of the metalloenzyme are represented by the two silvered spheres.

and Gram-negative bacteria. Genes encoding many variants of the enzyme are either chromosomally or plasmid-located. A major subtype in Gram-negative bacteria is found on transposon Tn9. CATs normally exist in solution as trimers with subunit molecular masses of between 24 and 26 kDa.

Catalytic process of chloramphenicol acetyl transferase

CAT metabolizes chloramphenicol in a two-stage process to a 1,3-diacetoxy derivative (Figure 9.5). The antibiotic is first converted to the 3-acetoxy compound using acetyl coenzyme A as an essential cofactor. A slow, nonenzymic rearrangement then transfers the acetoxy group to the 1 position. A second round of enzymic acetylation at the 3 position generates the final 1,3-diacetoxy product, although this reaction is much slower than the first step, owing to the impaired fit of the 1-acetoxychloramphenicol into the active site of the enzyme. Since both the mono- and diacetoxy derivatives are inactive as antibiotics, the two-stage acetylation sequence is biologically inefficient, but it is nevertheless a consequence of the spontaneous shift of the acetyl group from the 1 to the 3 position. Although the chloramphenicol molecule is potentially vulnerable to various forms of metabolic inactivation, including dehalogenation, reduction of the nitro group

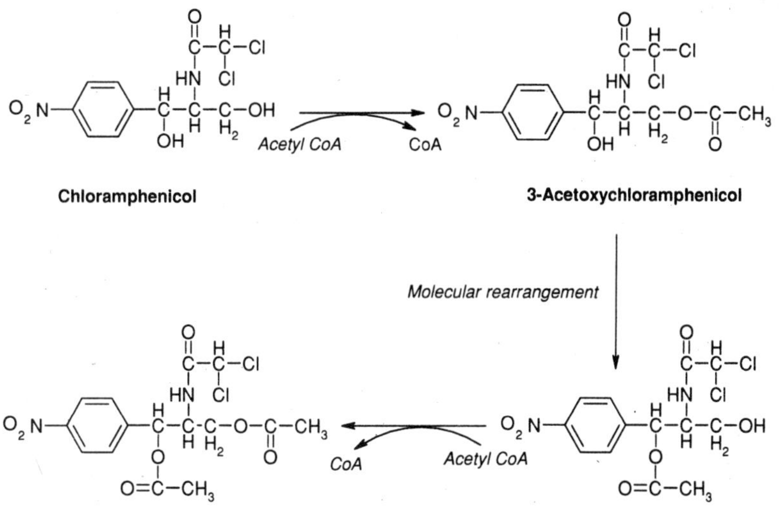

FIGURE 9.5 Inactivation of chloramphenicol by chloramphenicol acetyl transferase. 3-Acetoxychloramphenicol is formed first, followed by a nonenzymic shift of the acetyl group to the 1 position. A second enzymically catalyzed acetylation at the 3 position yields the 1,3-diacetoxy derivative.

and hydrolysis of the amide bond, the acetylation mechanism is the overwhelmingly significant contributor to resistance among bacterial pathogens.

Of the various forms of CAT defined by their structural and biochemical properties, the type III enzyme (CAT_{III}) has been most thoroughly studied. The determination of the structure of CAT_{III} by X-ray crystallography was a major contributor to the understanding of the catalytic mechanism (Figure 9.6). The trimeric holoenzyme has three identical active sites in the interfacial clefts between the monomers. Histidine-195 in one face of each cleft acts as a general base to abstract a proton from the 3-hydroxyl group of chloramphenicol. The resulting oxyanion attacks the 2-carbonyl carbon atom of acetyl coenzyme A to yield a tetrahedral intermediate. The oxygen atom of the intermediate is hydrogen bonded to the hydroxyl group of serine-148. The intermediate may also form hydrogen bonds with a water molecule linked to threonine-174. Finally, the tetrahedral intermediate collapses to yield 3-*O*-acetylchloramphenicol and free coenzyme A. The first-stage acetylation by CAT_{III} is an extremely efficient reaction, with a turnover rate of 600 molecules s^{-1}.

Physiology of chloramphenicol acetyl transferase synthesis

Gram-positive bacteria. The genes encoding CATs in Gram-positive bacteria such as *Staphylococcus* spp. and *Bacillus* spp. are inducible by chloramphenicol. However, the mechanism of induction does not involve increased transcriptional activity but rather an activation of the translation of the mRNA for the enzyme. Investigation of the inducible *cat* gene from *Bacillus pumilus* revealed that there is an 86-base pair region immediately 5′ to the coding sequence for the enzyme. Within the 86-base pair region, there are two distinct domains: domain A contains a ribosome-binding site (RBS-2), a translation initiation codon (GTG) and an open reading frame of nine codons ending in an upstream inverted-repeat sequence within domain B. The latter domain has two 14-base pair inverted repeat

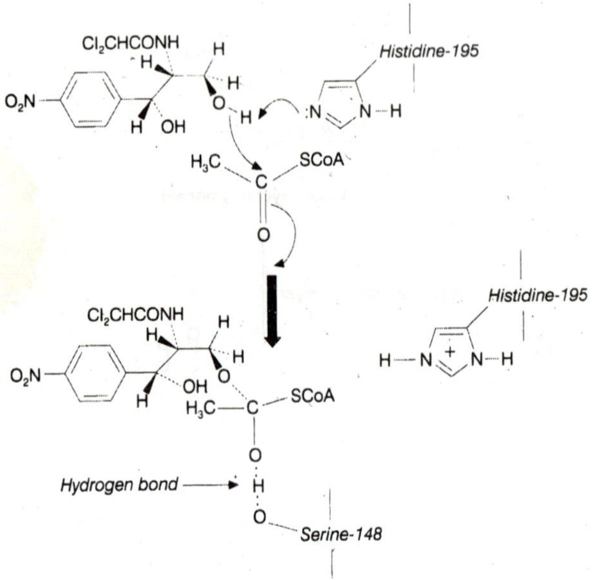

FIGURE 9.6 The key events in the catalytic mechanism of chloramphenicol acetyl transferase. Histidine-195 at each of the three active centres acts as a general base to abstract a proton from the 3-OH group of the antibiotic. The resulting activated oxygen attacks the 2-carbonyl group of acetyl coenzyme A to generate a tetrahedral intermediate that is hydrogen bonded to the γ-OH group of serine-148 of the enzyme. Subsequently the intermediate collapses to release 3-acetoxychloramphenicol, coenzyme A and free enzyme. [Reaction mechanism adapted with kind permission of I. A. Murray and W. V. Shaw and the American Society of Microbiology; *Antimicrob. Agents Chemother.* **41**, 1 (1997).]

sequences separated by 12 base pairs. The downstream inverted repeat spans the specific ribosome-binding site (RBS-3) for the *cat* gene transcript.

The sequence of domain B suggests that its transcript contains a stable stem loop whose secondary structure hinders translation of the *cat* gene transcripts. A detailed analysis of the transcriptional mechanism shows that while the *cat* coding sequence and the upstream 86-base pair region can be transcribed into a single mRNA molecule, about 50% of the observed transcripts terminate immediately after the regulatory region. This suggests that the mRNA stem loop also acts as a weak transcription termination signal. In the absence of chloramphenicol, even the full-length transcripts are not efficiently translated into enzyme protein because RBS-3 is hidden within the stem loop structure.

Addition of the antibiotic causes a ribosome engaged in translating the leading sequence of domain A to stall (remember that chloramphenicol inhibits protein synthesis, Chapter 5). The stalled, chloramphenicol-bound ribosome masks sequences in the mRNA, leading to destabilization of the stem loop secondary structure of domain B. The previously hidden RBS-3 is now made available to chloramphenicol-free ribosomes to initiate translation of the *cat* gene transcripts. Remarkably, therefore, the ability of chloramphenicol to inhibit protein biosynthesis facilitates the efficient translation of the mRNA and synthesis of the enzyme that inactivates the antibiotic. The principal result of the destabilization of the stem loop is to enhance the efficiency of translation of the *cat* gene transcripts, with only a small effect in relieving transcription termination. This mechanism of induction is referred to as translational attenuation, and we shall see it in action again in inducible resistance to erythromycin and tetracyclines.

Gram-negative bacteria. CAT synthesis is constitutive in most Gram-negative bacteria. However, CAT synthesis in *Escherichia coli* is subject to catabolite repression. Synthesis is faster in cultures grown on glycerol than in glucose-supported cultures. Cyclic AMP complexed with the catabolite activator protein (CAP) is required for optimal CAT synthesis. During glucose-supported growth, the intracellular levels of cyclic AMP are low and the uncomplexed CAP cannot bind to the promoter region that regulates the *cat* gene.

Transcription of the *cat* gene is therefore inefficient. Conversely, when cyclic AMP levels rise, the CAP-cyclic AMP complex binds to the promoter region, facilitating the interaction of DNA-dependent RNA polymerase with the structural gene and its transcription.

9.1.3 Aminoglycosides

Bacterial resistance to aminoglycosides can result from mutations affecting the ribosomes and from changes in cellular permeability, but the most important cause of resistance is enzymically catalyzed inactivation of the antibiotics. Although more than 50 aminoglycoside-inactivating enzymes have been reported, they catalyze only three major types of reactions:

1. *N*-Acetylation of vulnerable amino groups using acetyl coenzyme A as the acetyl donor. The *N*-acetyl transferases comprise the largest group of aminoglycoside-inactivating enzymes.
2. *O*-Adenylylation involving the transfer of an AMP residue from ATP to certain hydroxyl groups. The *O*-adenylyl transferases form the smallest group of inactivating enzymes.
3. *O*-Phosphorylation of hydroxyl groups with ATP acting as the phosphate donor.

The enzymes are further divided into subgroups according to the molecular position on the drugs at which these reactions occur. This results in a very complex array of bacterial defences against the aminoglycosides. There are now examples of X-ray-determined three-dimensional structures from each of the three groups of aminoglycoside-inactivating enzymes.

Typical reactions involving streptomycin and kanamycin A are shown in Figure 9.7. Streptomycin is subject to both adenylylation and phosphorylation but

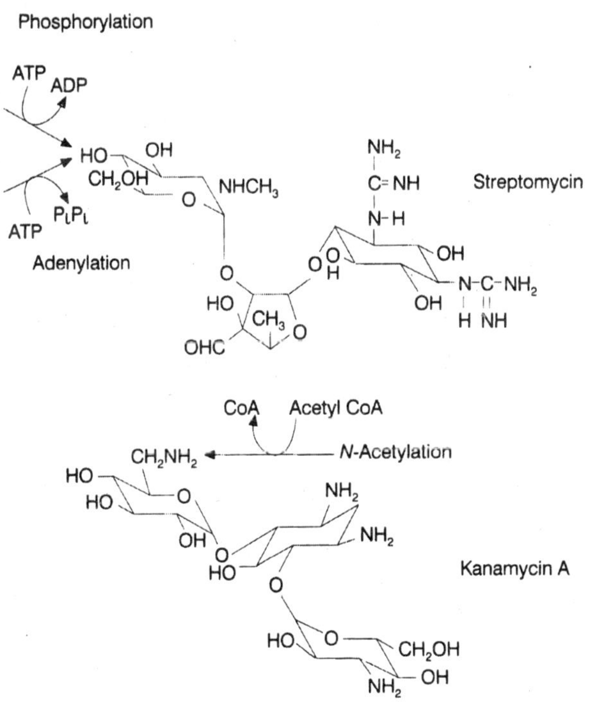

FIGURE 9.7 Three modes of enzymic inactivation of aminoglycoside antibiotics. Unlike kanamycin A, streptomycin is not subject to *N*-acetylation, while kanamycin A is also inactivated by *O*-adenylylation and *O*-phosphorylation. An extensive array of bacterial enzymes is involved in the inactivation of aminoglycosides. A review listed at the end of the chapter provides further details.

it is not a substrate for the *N*-acetyl transferases. The *N*-acetyl transferases are specific for the amino groups of other aminoglycosides in four different positions (1, 3, 6′ and 2′). The *O*-adenylyl transferases attack OH groups in 2″, 3″ and 4 positions and the *O*-phosphoryl transferases target OH groups in the 3′, 3″ and 4 positions. Isozymes exist for many of these enzymes. A bifunctional enzyme that catalyzes both *N*-acetyl and *O*-phosphoryl transferase activities appears to be the result of gene fusion, with protein domains responsible for each enzymic activity being derived from different genes. For a comprehensive account of the many enzymes that metabolize aminoglycosides, the reader is referred to a review in 'Further reading.'

The cellular location of the aminoglycoside-modifying enzymes is somewhat uncertain. Because the target of aminoglycoside action is at the ribosome, the cytoplasm was thought to be the most likely location for the inactivating enzymes. However, a cytoplasmic location might be relatively inefficient in protecting the ribosomes because the enzymes are only synthesized in small amounts. It would make more biological 'sense', therefore, if the enzymes were secreted into the periplasm where aminoglycoside inactivation would occur before entering the cytoplasm. However, it not clear how the essential cosubstrates ATP and acetyl coenzyme A could gain access to the periplasm. Nevertheless, there is evidence for the location of at least one *O*-adenylyl transferase in the periplasm. Signal sequences of 20–30 amino acids at the N-terminals of proteins ensure the export of proteins across the cytoplasmic membrane into the periplasm. Many members of the *N*-acetyl transferase family have these sequences, although they are not found among the *O*-phosphoryl transferases. Thus although the periplasm appears to be an optimum location for aminoglycoside-modifying enzymes, this may not always be the case. In most bacteria the expression of genes encoding aminoglycoside-modifying enzymes is not subject to regulation. However, the genes for *N*-acetyltransferases in *Serratia marcescens* and *Providencia stuarti* appear to be tightly regulated, although in some clinical isolates of aminoglycoside-resistant strains the control of expression is relaxed.

Surveys of aminoglycoside resistance among bacterial pathogens from countries around the world reveal a complex pattern in which many organisms harbour a combination of several genes, either chromosomally or plasmid-located, expressing different modes of aminoglycoside metabolism. Resistance to a range of aminoglycosides is especially marked among *Citrobacter, Enterobacter* and *Klebsiella* species, although other Gram-negative bacteria also exhibit resistance. Several semisynthetic aminoglycosides have been designed to counter the action of the inactivating enzymes. Tobramycin, netilmicin and amikacin were all developed to resist *O*-phosphotransferases but were nevertheless found to be susceptible to *N*-acetyl transferases. The range and diversity of aminoglycoside-inactivating enzymes will probably continue to frustrate efforts to devise compounds that resist deactivation. An alternative possibility lies with enzyme inhibitors, although the discovery of agents with broad specificity would be challenging. Specific inhibition of the *O*-phosphotransferases is potentially attractive because of some structural similarity between these enzymes and the eukaryotic protein phosphorylating kinases. The latter enzymes are of great interest in mammalian pharmacology and many inhibitors are already available, some of which show a degree of activity against the aminoglycoside *O*-phosphotransferases.

9.1.4 Streptogramins

The ever-growing threat of bacterial resistance to the most commonly used antibiotics has encouraged the introduction of a limited number of more unusual drugs into human clinical medicine, including the streptogramins. As described in Chapter 5, group A compounds such as dalfopristin are given in combination with group B compounds like quinupristin to ensure a synergistic antibacterial effect. This combination is effective against dangerous pathogens, including multidrug resistant *Staphylococcus aureus* and vancomycin-resistant *Enterococcus faecium*. A commonly encountered form of resistance to the group A compounds is due to *O*-acetylation of the lone hydroxyl group on these drugs (e.g. dalfopristin, Figure 5.14). The reaction is catalyzed by an *O*-acetyl transferase encoded by a family of plasmid-borne *vat* genes in Gram-positive cocci. This enzyme shares some amino acid sequence similarity with acetyl transferases that confer low-level resistance to chloramphenicol. However, the latter enzymes should not be confused with the 'classic' CAT_{III} *O*-acetyl transferase described previously, which is responsible for high-level resistance to chloramphenicol. X-Ray analysis of crystals of the dalfopristin *O*-acetyl transferase reveals it as a homotrimeric enzyme with the active center located between two adjacent subunits. The acetyl donor is acetyl coenzyme A. Details of the complex structure of the enzyme and its interactions with the antibiotic and acetyl coenzyme A are available in a reference listed in 'Further reading.'

The appearance of human pathogenic cocci resistance to group A streptogramins may be linked to the widespread use of the group A compound virginiamycin as a growth promoter in livestock since the *vatA* gene mediates resistance to both dalfopristin and virginiamycin. A detailed understanding of the structure of the *O*-acetyl transferase and its interactions with antibiotic substrates and the cofactor may assist in the design of novel group A compounds which are impervious to *O*-acetylation.

A lyase enzyme has been isolated from *Staphylococcus aureus* that inactivates group B streptogramins by cleaving the hexadepsipeptide chain of the antibiotics.

9.2 Loss or downregulation of drug activation

9.2.1 Metronidazole

As described in Chapter 6, this drug has important applications in the treatment of infections caused by strictly anaerobic bacterial pathogens and certain protozoal parasites and also in the elimination of the causative organism of peptic ulcer disease, the microaerophilic bacterium *Helicobacter pylori*. The activity of metronidazole depends upon its metabolic reduction by the target pathogens to various short-lived, highly reactive radicals which attack DNA and susceptible proteins. Resistance to metronidazole occurs in all of the target species and generally appears to be due to loss or downregulation of the enzymes involved in the activation of the drug. In susceptible *Helicobacter pylori*, activation is carried out by a nitro reductase encoded by the *rdxA* gene. Metronidazole resistance in most clinical isolates is associated with null mutations in the *rdxA* gene which prevent the reductive activation of the drug. Resistance in the parasitic protozoa *Giardia duodenalis* and *Trichomonas vaginalis* is due to downregulation of the pyruvate:ferredoxin oxido reductase required for the activation of metronidazole in these species.

9.2.2 Isoniazid

The efficacy of isoniazid against *Mycobacterium tuberculosis* depends initially on the intracellular activation of the compound by a bacterial catalase-peroxidase encoded by the *katG* gene (Chapter 2). While the identity of the active metabolite of isoniazid is still a matter of some debate, it is clear that mutations adversely affecting *katG* confer bacterial resistance. Mutations in the *katG* gene are the major cause of high-level resistance to isoniazid in clinical isolates of *Mycobacterium tuberculosis* from around the world. Of these mutations, the replacement of serine at position 315 of the catalase-peroxidase by threonine is one of the most frequently encountered. This mutation results in a markedly reduced enzyme affinity for isoniazid, so that there is little catalase-peroxidase-mediated conversion of the drug to the active molecular species. Although the mutant enzyme has somewhat lower catalytic activity for its physiological substrates than the wild-type enzyme, the virulence of *Mycobacterium tuberculosis* expressing the mutant enzyme is apparently unaffected. An understanding of the nature of the binding interaction between isoniazid and the wild-type enzyme and how this is impaired by the threonine-for-serine mutation may useful in the design of isoniazid derivatives that retain substrate activity for the mutant enzyme.

9.2.3 Pyrazinamide

This antitubercular drug also requires activation from the inactive parent compound to the inhibitory pyrazinoic acid by a mycobacterial enzyme referred to as pyrazinamidase (Chapter 2). Some clinical isolates of

Mycobacterium tuberculosis lack this enzyme and are resistant to pyraziamide as a result.

9.3 Modification of drug targets

9.3.1 β-Lactams

While the most common mechanism of resistance to β-lactam antibiotics is that of inactivation by β-lactamases, resistance can result from amino acid changes in penicillin-binding proteins that depress their affinity for β-lactams. Mosaic genes encoding hybrid PBPs with reduced affinity for β-lactams were described in Chapter 8. Such genes confer resistance to both penicillins and cephalosporins in meningococci, gonococci and pneumococci. In this section we concentrate on another example of β-lactam resistance caused by a target change that is causing grave concern: the resistance of *Staphylococcus aureus* to methicillin.

This organism was rightly regarded as a particularly dangerous pathogen in the preantibiotic era. The introduction of benzylpenicillin created a fortunate interlude during which staphylococcal infections responded readily to the new drug. The subsequent rise of β-lactamase-mediated resistance in staphylococci was initially countered by the introduction of the semisynthetic, β-lactamase-stable methicillin. However, the 1980s saw the emergence of methicillin-resistant *Staphylococcus aureus* (MRSA), which is also resistant to all other β-lactams. Resistance is caused by the acquisition of a transposon-located novel gene, *mecA*, that encodes a novel PBP (PBP2′, otherwise designated as PBP2a) with very low affinity for all β-lactams. The transposon integrates into the chromosome of *Staphylococcus aureus* to create a 50-kb resistance island which is widely distributed among staphylococcal species. Remarkably, the PBP2′ protein takes over the function of all the other PBPs, thus rendering the growth of *mecA*$^+$ bacteria resistant to methicillin and other β-lactams.

Some bacterial strains have an upstream regulatory region, *mecR1-mecI*, that negatively controls the expression of *mecA*. In the absence of methicillin, synthesis of the MecR1 protein is repressed. Resistance is slowly induced to low levels by methicillin. Mutations in, or complete loss of, the *mecI* region permit the synthesis of both MecR1 and PBP2′, giving rise to high levels of methicillin resistance. There is marked sequence homology of *mecR1* and *mecI* with the *blaR1* and *blaI* genes that regulate β-lactamase synthesis in Gram-positive bacteria. Consequently the *mecA* gene is also regulated by *blaR1* and *blaI* and PBP2′ expression is induced by β-lactams that are recognized by the BlaR1 protein.

Although PBP2′ replaces the function of the other PBPs in *Staphylococcus aureus*, its precise function in peptidoglycan synthesis is not clear and its activity leads to peptidoglycan with a lower than normal degree of cross-linking. The level of PBP2′ synthesis in *mecA*$^+$ bacteria does not correlate closely with the level of resistance, and it is known that other genes contribute to methicillin resistance, possibly even including genes that encode 'super' penicillinases capable of degrading methicillin.

9.3.2 Macrolides

The inhibition of protein synthesis by erythromycin and other macrolides depends upon their interaction with domain V of 23S rRNA (Chapter 5). While there are bacterial strains that enzymically inactivate erythromycin and other macrolide antibiotics, macrolide metabolism does not contribute significantly to the problem of clinical resistance. Most macrolide-resistant Gram-positive pathogens, including *Staphylococcus aureus* and *Streptococcus* spp., harbour plasmid-borne genes (*erm*) that encode a family of *N*-methyl transferases, or methylases as they are also known. The substrates for these enzymes are specific adenine residues in 23S rRNA in the domain involved in the interaction of erythromycin with the ribosome. There are no N^6-methylated adenine residues in the 23S rRNA of wild-type, erythromycin-sensitive bacteria. In contrast, N^6-dimethylated adenine appears in the 23S rRNA of resistant *ermA*$^+$ *Staphylococcus aureus* cells grown in medium containing erythromycin. Evidence that the modified 23S rRNA confers ribosomal resistance is provided by the observation that 70S ribosomes reconstituted with 23S rRNA from resistant *Bacillus subtilis* and ribosomal protein from erythromycin-susceptible cells are resistant to erythromycin. The critical adenine residues targeted in the peptidyl transferase domain by the *erm*-mediated methylase are A-2058 in *Escherichia coli* and A-2086 in *Bacillus stearothermophilus*. Dimethylation prevents the necessary hydrogen bonding between these adenine residues and the 2′ hydroxyl group of the desosamine ring of the macrolides and sterically hinders antibiotic access to the inhibitory binding site.

The *N*-methylase products of the *erm* gene family use *S*-adenosylmethione as the methyl donor. Some of the enzymes transfer a single methyl group to the N^6 position of adenine, whereas others donate two groups. Dimethylases may confer a higher level of resistance to a broader range of macrolide antibiotics than the monomethylases. The 23S rRNA substrate is presented to the methylases as a component of nascent ribosomes rather than as part of the mature, functioning particles.

Regulation of *erm* gene expression

The expression of *erm* genes is inducible by erythromycin. The inducing activity of erythromycin is closely linked to its ability to inhibit ribosomal function, and erythromycin derivatives devoid of inhibitory activity cannot induce expression of the *N*-methylases. The mechanism of erythromycin induction of the *N*-methylases is analogous to that of chloramphenicol acetyl transferase by chloramphenicol, i.e. induction is achieved by translational attenuation rather than increased gene transcription. Translation of the *erm* mRNA is slow and inefficient in the absence of erythromycin because of the unfavourable conformation of a 141 nucleotide leader sequence upstream of the open reading frame for the enzyme. The secondary structure of the leader sequence is believed to mask the first two codons of the orf as well as the ribosomal binding site for the mRNA, resulting in very low constitutive synthesis of *N*-methylase. However, when translation of the leader sequence is inhibited by binding of erythromycin to ribosomes, the efficiency of subsequent translation is increased, probably because the secondary structure of the leader sequence undergoes a conformational rearrangement to a new state that is consistent with a more rapid readthrough of the orf and increased enzyme synthesis. Mutations in the *erm* genes that destabilize the secondary structure of the leader sequences result in higher levels of constitutive *N*-methylase biosynthesis.

9.3.3 Quinolones

The role of the quinolones in the treatment of bacterial infections has grown steadily with the introduction of novel derivatives with broader spectra of action than the progenitor compound, nalidixic acid. The emergence of resistance accompanying this increasing use of quinolones is due largely to mutations in the A (GyrA) subunit of the target enzyme, topoisomerase II, or DNA gyrase (Chapter 4), although drug efflux may also contribute to the problem. Enzymic inactivation of quinolones has not so far been detected as a mechanism of bacterial resistance.

The GyrA subunits of the tetrameric enzyme ($GyrA_2GyrB_2$) are responsible for introducing double-stranded breaks in DNA and for subsequent resealing of the breaks during the negative supercoiling process. A region described as the quinolone-resistance-determining region (QRDR) is located between amino acid residues 67 and 107 of GyrA in *Escherichia coli*. This sequence of amino acids gives rise to a positively charged molecular surface to which the DNA is believed to bind. Furthermore, the region around serine-83 is critical for the enzyme-quinolone interaction and in several quinolone-resistant strains of *Escherichia coli* this amino acid is substituted by nonpolar, bulkier amino acids such as leucine, alanine or tryptophan. In *Staphylococcus aureus*, the homologous position is serine-84, which is replaced by leucine in many quinolone-resistant clinical isolates. Quinolone-resistant *Staphylococcus aureus* and other Gram-positive pathogens may also have mutations in the ParC subunit of topoisomerase IV (Chapter 5), strengthening the supposition that this enzyme is also a significant target for quinolones in Gram-positive species. In the GyrA subunit of *Campylobacter jejuni*, *Klebsiella pneumoniae* and *Pseudomonas aeruginosa*, threonine is the normal amino acid at position 83 instead of serine and these organisms are intrinsically much more resistant to quinolones. The greater bulk of threonine at position 83 probably hinders the optimal binding of quinolone to the target site.

9.3.4 Streptomycin

In addition to inactivation by aminoglycoside-modifying enzymes, resistance to streptomycin also arises from mutations affecting the target site in the 30S ribosomal subunit (Chapter 5). Streptomycin still finds some application in the treatment of tuberculosis. DNA analysis of clinical isolates of resistant *Mycobacterium tuberculosis* shows that about 70% have mutations affecting the *rpsL* gene that codes for protein S12 which contributes to the binding of streptomycin to the ribosomal subunit. The mutations cause replacement of lysine-43 or lysine-88 by arginine and a consequent loss of binding of streptomycin to the ribosome. Mutations affecting the highly conserved position 904 of 16S rRNA also cause streptomycin resistance in some *Mycobacterium tuberculosis* isolates. Ribosomal resistance to streptomycin is found in clinical isolates of other bacteria, including *Neisseria gonorrhoeae, Staphylococcus aureus* and *Streptococcus faecalis*.

A recently discovered novel mechanism for the protection of ribosomes against several aminoglycoside antibiotics, except streptomycin, is a plasmid-mediated enzyme which methylates a residue in the A site of 16S rRNA, probably guanine-1405. This form of ribosomal protection against valuable drugs such as tobramycin and amikacin has appeared in clinical isolates of *Pseudomonas aeruginosa* and *Serratia marcescens*.

9.3.5 Rifampicin

Resistance to rifampicin is proving to be a considerable threat to the successful treatment of tuberculosis. Originally this drug was highly effective against *Mycobacterium tuberculosis*, but mutations affecting the β subunit of the target enzyme, DNA-dependent RNA polymerase (Chapter 4), are often responsible for the loss of bacterial sensitivity to rifampicin. Most resistant clinical isolates of *Mycobacterium tuberculosis* have replacements at serine-531 or histidine-526. These changes, caused by mutations near the center of the *rpoB* gene which encodes the β subunit, are readily detected by DNA analysis and provide a rapid test for drug resistance. A detailed analysis of a range of rifampicin-resistance mutations in *Escherichia coli* and *Mycobacterium tuberculosis* indicates that the amino acid substitutions either directly or indirectly adversely affect the interaction of rifampicin with its binding site, thus reducing the affinity of the antibiotic for its target enzyme. Mutations conferring rifampicin resistance do not affect the growth of *Mycobacterium tuberculosis*, whereas rifampicin resistance in *Escherichia coli* causes slower growth. The reason for this difference is not known, but it may be associated with the naturally slower growth of *Mycobacterium tuberculosis* compared with that of *Escherichia coli*.

9.3.6 Inhibitors of dihydrofolate reductase

Trimethoprim

The basis of one major form of resistance to trimethoprim is analogous to that of methicillin resistance; namely, resistant cells acquire additional genetic information for a molecular target with reduced susceptibility to the drug. In Gram-negative bacteria, 16 different genes for trimethoprim-resistant DHFRs have been identified. These genes are mostly plasmid-borne, although they may temporarily reside on the chromosome because of their association with transposons. The enzymes fall into two families. Family 1 has five members with polypeptide chains sharing 64–88% sequence identity. The enzymes of family 1 are homodimeric proteins with IC_{50} values between 1 and 100 μM, compared with an IC_{50} for the wild-type DHFR of 1n M. Family 2 is larger, with 11 members more closely related than those of family 1, with sequence identities of 78–86%. The enzymes of family 2 are all homotetramers and are highly resistant to trimethoprim, with IC_{50} values greater than 1 mM. The most widely distributed trimethoprim-resistant DHFR amongst Gram-negative bacteria is encoded by the *dhfrI* gene and belongs to family 1. The *dhfrI* gene is located within a highly mobile cassette associated with a promiscuous transposon (Tn7) that inserts into the chromosomes of many bacteria.

Some trimethoprim-resistant bacteria overproduce modified DHFRs. For example, a highly resistant strain of *Escherichia coli* overproduces by about 100-fold an enzyme that is about threefold more resistant to trimethoprim than the wild-type enzyme. Trimethoprim-resistant isolates of *Haemophilus influenzae* (an important clinical target) also overproduce DHFR, although in this case the enzyme is 100- to 300-fold more resistant to the drug. Enzyme overproduction in both species of bacteria is associated with mutations in the promoter sequences that control expression of the structural genes.

It is interesting that the kinetic parameters of trimethoprim-resistant enzymes for the normal substrates, dihydrofolate and NADPH, are essentially unchanged. The mutant enzymes are therefore competent to take over the metabolic functions of the drug-sensitive enzyme.

Sulfonamides

Trimethoprim is usually administered in combination with a sulfonamide such as sulfamethoxazole. Unfortunately, bacterial resistance to sulfonamides often coexists with trimethoprim resistance. Sulfonamide resistance can be due either to mutations in the chromosomal gene, *dhps*, that mediates dihydropteroate synthase, or to the acquisition of plasmid-borne genes coding for sulfonamide-resistant forms of the enzyme. In *Neisseria meningitidis* there are two variants of a chromosomally mediated resistant dihydropteroate synthase, one of which may be the result of recombination between two related *dhps* genes. At least two types of plasmid-borne *dhps* genes code for sulfonamide-resistant enzymes in Gram-negative bacteria. In all these resistant enzymes the Michaelis constants (K_m) for the natural substrate, *p*-aminobenzoic acid, are similar to those for the drug-sensitive enzyme. Sulfonamide-resistant forms of dihydropteroate synthase have also been found in strains of the malarial parasite *Plasmodium falciparum* in regions of the world where the sulfa drug sulfadoxime is used in combination with inhibitors of dihydrofolate reductase.

Pyrimethamine and cycloguanil

Inhibitors of DHFR, including pyrimethamine and the liver metabolite of proguanil, cycloguanil (Chapter 6), have been central to the prophylaxis and treatment of malaria for more than 60 years. The relentless increase in resistance to these drugs in many parts of the world is therefore a major threat to the containment of one of the most prevalent infections on Earth. Because of the many technical difficulties in working with protozoal parasites, the definition of the mechanisms of resistance in naturally occurring infections has been extremely difficult. Much of the available biochemical information has therefore been obtained with drug-resistant malarial protozoa developed in the laboratory. Nevertheless it is believed that this information gives a reasonable indication of the nature of drug resistance in the 'field'. Furthermore, by using technologies such as the polymerase chain reaction for genetic analysis, it is now possible to correlate laboratory and field studies of drug-resistant protozoa.

Resistance to pyrimethamine in the most dangerous malarial parasite, *Plasmodium falciparum*, is commonly due to a single-point mutation in DHFR that replaces serine-108 with asparagine. This mutation has been detected in laboratory isolates and from patients, but surprisingly it does not confer cross-resistance to the structurally similar cycloguanil. However, when serine-108 is replaced by threonine together with a valine-for-alanine replacement at position 16, the parasite becomes resistant to cycloguanil but not to pyrimethamine. Cross-resistance to both drugs occurs in strains harboring a combination of mutations at positions 51, 57, 108 and 164 which arise from genetic recombination during the sexual reproduction phase of the parasite's life cycle.

A three-dimensional model of the DHFR from *Plasmodium falciparum* based on its homology with other DHFRs was used to explore the effects of the mutations at positions 16 and 108 by virtual 'docking' experiments in an attempt to understand the lack of cross-resistance to cycloguanil and pyrimethamine in such mutants. Pyrimethamine was found to dock equally well with both the double-mutant enzyme and the wild-type enzyme, whereas the binding of cycloguanil was sterically hindered by the bulky valine residue at position 16 in the mutant enzyme. While the binding of inhibitors was progressively weakened by the gradual accumulation of mutations in DHFR, the catalytic efficiency of the enzyme remained unaffected in virulent strains of the parasite. Certain strains of *Plasmodium falciparum* and another protozoal pathogen, *Leishmania major*, resist pyrimethamine by overproducing DHFR, either by gene duplication or by increasing expression levels.

9.3.7 Inhibitors of HIV reverse transcriptase

Nucleoside analogues

The introduction of AZT (3′-azido-3′-deoxythymidine, Chapter 4) was a therapeutic landmark in the treatment of AIDS. AZT has been followed by other nucleoside analogue inhibitors of the reverse transcriptase (RT) of HIV and by non-nucleoside inhibitors such as nevirapine. Unfortunately, the remarkable propensity of HIV for mutation (Chapter 8), combined with the inability of RT inhibitors to suppress viral replication by more than 90%, leads to the rapid emergence of drug-resistant strains of the virus. Resistance to nucleoside inhibitors of RT is often due to mutations close to the nucleotide binding site which map to the β3–β4 loop in the fingers of the p66 subunit (Chapter 4). High-level resistance to lamivudine is caused by a valine-for-methionine substitution at position 184, which is close to the ribose ring of the nucleoside triphosphate substrate. The increased bulk of the valine residue stericially hinders the oxathiolane ring of lamivudine, thus reducing the binding affinity of the enzyme for the triphosphate metabolite of the drug. A group of mutations referred to as the Q151M complex causes progressively increasing resistance to the nucleoside inhibitors of RT. Usually the first mutation replaces glutamine (Q) at position 151 with methionine (M), which is situated very close to the nucleotide binding site. This mutation is followed by a series of secondary mutations which gradually increase the activity of the enzyme and enhance viral drug resistance.

Another interesting resistance mutation of RT results in the removal of the incorporated terminal nucleoside analogue from the viral DNA chain, thereby allowing the normal extension and completion of the viral genome. Apparently the mutated enzyme promotes the incorporation of either ATP or pyrophosphate into the DNA chain adjacent to the incorporated analogue. In this position the ATP or pyrophosphate can attack the phosphodiester bond linking the analogue to the DNA chain, resulting in the expulsion of the drug from its terminal position. Mutations of this type occur at residues distant from the nucleotide binding site, e.g. leucine-for-methionine at position 41 and tryptophan-for-leucine at position 210.

Non-nucleoside inhibitors of HIV reverse transcriptase

In contrast with the nucleoside analogues, the NNRTIs, including such compounds as nevirapine and efavirenz (Chapter 4), are noncompetitive inhibitors of RT. The NNRTIs bind to a hydrophobic pocket some 10Å from the catalytic center of the enzyme. Mutations causing resistance to nevirapine and efavirenz are located within two β-strands in the NNRTI-binding pocket between residues 100–110 and 180–190 and reduce the affinity between enzyme and drug. Fortunately, therefore, these mutations do not cause cross-resistance to the nucleoside analogues, although cross-resistance can occur amongst different NNRTIs.

9.3.8 Inhibitors of HIV protease

These drugs (Chapter 6) form part of the standard triple therapy for AIDS in combination with two different inhibitors of RT. Triple therapy is considered essential because mutant viruses with drug-resistant proteases emerge readily when protease inhibitors are given as a single therapy. It may be recalled that the antiviral effectiveness of protease inhibitors is due to their suppression of the cleavage of the precursor viral polyproteins, which is an essential stage in the cycle of viral replication. The development of resistance during monotherapy with the protease inhibitor ritonavir (Chapter 4) has been documented in considerable detail. The gradual loss of effectiveness of the drug in AIDS patients is associated with a sequential accumulation of mutations in the target enzyme. The rate at which mutations appear is inversely related to the concentration of drug in the patient's blood, strongly suggesting that blood levels of the drug should be maintained as high as possible to minimize viral replication. Mutations at single loci in HIV protease are associated with low-level viral resistance to ritonavir: a 7- to 10-fold increase in resistance requires the accumulation of three to four mutations, and high-level resistance (> 20-fold) requires four to five mutations. Of the nine amino acid changes contributing significantly to resistance, replacement of valine-82 is probably the most important. X-Ray analysis of the susceptible enzyme shows that this residue interacts directly with ritonavir.

Ritonavir-resistant viruses are only partially cross-resistant to other protease inhibitors, indicating that there are significant and clinically important differences in the molecular interactions of the various inhibitors with the mutant enzymes.

Because mutations conferring resistance to inhibitors of HIV protease are usually located in the substrate binding pocket, they may also reduce the catalytic efficiency of the enzyme. However, these adverse effects are often compensated by mutations elswhere in the enzyme which enhance its stability or catalytic activity, thus helping to restore the proliferative activity of the virus. Resistance to protease inhibitors is also associated with mutations in the viral polyprotein substrate for the protease which increase its susceptibility to cleavage by the enzyme.

9.3.9 Acyclovir

Infections caused by the various herpes viruses range from the relatively trivial to severely disabling or even life-threatening. One of the most important applications of acyclovir (Chapter 4) is the treatment of herpes infections in immunosuppressed patients, in whom drug-resistant forms of the viruses develop most readily. Acyclovir is a prodrug that is first converted to the monophosphate derivative by thymidine kinase (TK) encoded in the viral genome, and subsequently to the inhibitory triphosphate by enzymes of the infected host cell. One relatively uncommon form of resistance to acyclovir results from mutations affecting the viral DNA polymerase which is the ultimate target for the drug. The modified enzyme has a diminished affinity for acyclovir triphosphate while retaining a relatively unchanged ability to bind the four nucleoside triphosphates required for viral DNA synthesis. However, by far the most common mechanism of resistance to acyclovir in immunosuppressed patients depends on mutations in the viral TK. The acquisition of an inappropriately placed stop codon results in a truncated polypeptide with little or no enzymic activity. Viruses lacking TK activity cannot phosphorylate acyclovir and lose the characteristic ability of herpes viruses to reactivate from a dormant state in neuronal cells. Missense mutations in viral TK, on the other hand, allow the enzyme to retain its affinity for thymidine while markedly depressing that for acyclovir and, furthermore, do not affect the reactivational capability of the virus.

9.3.10 Antiinfluenza inhibitors of viral neuraminidase

Since the introduction of zanamivir and oseltamivir between 1999 and 2002 for the treatment of influenza (Chapter 6), there has been extensive surveillance of clinical isolates in order to assess the likelihood of the emergence of drug-resistant viruses. A laboratory-derived mutation with a valine-for-glutamic acid replacement at position 119 in viral neuraminidase, which causes the loss of enzyme affinity for both zanamivir and oseltamivir, has also been observed in viruses isolated from patients treated with oseltamivir. However, although resistant viruses with mutations in the enzyme target are readily generated in the laboratory by repeated passage of cultured virus in the presence of drugs, the overall incidence of comparable resistant strains from clinical isolates has so far been very low.

9.4 Drug efflux pumps

The various types of drug efflux pumps in prokaryotes and eukaryotes and their role in the intrinsic resistance of micro-organisms to drugs and other toxic chemicals were discussed in Chapter 7. This section is concerned with the part that the pumps play in acquired resistance to medically important antimicrobial drugs.

9.4.1 Tetracyclines

The clinical value of tetracyclines, which are among the cheapest and mostly widely used antibacterial drugs, has been severely compromised by the emergence of resistant bacteria in both Gram-positive and Gram-negative groups. There are two major mechanisms of resistance to tetracyclines: drug efflux and ribosomal protection. This account is concerned with tetracycline efflux; ribosomal protection is described in a later section.

The genes for tetracycline-specific, efflux-mediated resistance are almost exclusively plasmid and transposon-located, although there is also a chromosomal tetracycline efflux system in *Escherichia coli* associated with the global regulator locus, *marA*, that enhances intrinsic resistance to many drugs (Chapter 8). High-level expression of *marA* boosts the production of the multidrug efflux pump found in many wild-type Gram-negative bacteria, which includes tetracycline among its many substrates.

Some 25 genes currently identified among Gram-positive and Gram-negative bacteria encode tetracycline efflux pumps. Eighteen of them are designated as *tet* genes and seven as *otr* (oxytetracycline resistance), although there is no functional difference between *tet* and *otr* genes. New tetracycline efflux genes continue to be discovered. A review listed at the end of this chapter provides a comprehensive analysis of this complex topic. Most of the efflux pumps confer resistance to all tetracyclines except minocycline and glycylcyclines like tigicycline (Chapter 5).

The tetracycline efflux genes encode membrane-bound proteins of approximately 46 kDa, all belonging to the MFS family of efflux pumps (Chapter 7). The efflux proteins can be assigned to six groups according to the degree of shared amino acid sequence identity. In general, the Gram-negative proteins have 12 hydrophobic membrane-spanning domains and the Gram-positive proteins have 14. Regions of hydrophilic amino acids loop out into both the periplasmic and cytoplasmic regions. The tetracycline efflux proteins (Tet), which probably exist as multimers within the membrane, extrude a tetracycline molecule complexed with a divalent ion (probably Mg^{2+}) in exchange for a proton. Tetracycline is pumped out of the cytoplasm against its concentration gradient and the energy required for this is derived from the proton motive force. Mutational studies suggest that the cytoplasmic loops interact with the proton motive force, although the structural details are not known. There is no direct link between tetracycline efflux and ATP hydrolysis. Gram-negative Tet proteins have functional α and β domains, corresponding to the N- and C-terminal halves of the proteins. Genetic evidence indicates that amino acids distributed across both domains participate in the efflux function. Although the Tet proteins of Gram-negative bacteria are more closely related to each other than to the larger 14-transmembrane-domain proteins of Gram-positive bacteria, the conservation of certain sequence motifs across the species suggests that the mechanism of all tetracycline pumps driven by proton motive force is basically similar. The binding site for the transported tetracyclines appears to be located in helix 4 of the transmembrane domain. The tetracycline efflux proteins have both amino acid sequence and structural similarities with efflux pumps that extrude other drugs, including chloramphenicol and the quinolones as well as quaternary ammonium antiseptics.

Regulation of tetracycline resistance. It was discovered many years ago that efflux-dependent tetracycline resistance is inducible by tetracyclines in both Gram-negative and Gram-positive bacteria. A low level of resistance is evident when the cells are first exposed to antibiotic, followed by a rapid shift to high-level resistance. The mechanism of induction is different in Gram-negative and Gram-positive bacteria. The tetracycline efflux system in Gram-negative bacteria is mediated by a structural gene for the Tet protein and by a gene coding for a repressor protein. The two genes are arranged in opposite directions and share a central regulatory region with overlapping promoters and operators. In the absence of tetracycline, α-helices in the N-terminal domain of the repressor protein bind to the operator regions of the repressor and structural genes, thereby blocking the transcription of both. The introduction of tetracycline complexed with Mg^{2+} leads to binding of the drug to the repressor protein. This causes a conformational change in the repressor (which has been revealed by X-ray crystallography) that eliminates its binding to the operator region and permits transcription of the repressor and structural genes. The shift from low-level to high-level resistance takes place within minutes of exposure of the bacteria to tetracycline. The process is reversible and a downshift of resistance follows the removal of tetracycline from the bacterial environment.

In contrast to Gram-negative bacteria, there is no Tet repressor protein in Gram-positive cells. The regulation of tetracycline resistance in the Gram-positive bacteria examined so far involves another example of translational attenuation. The mRNA for the Tet protein has two ribosomal binding sites, RBS1 and RBS2. In the uninduced state, the ribosome binds to RBS1 and a short leader peptide sequence is translated before the RBS2 site, which precedes the start of the structural gene proper. At this stage the RBS2 site is thought to be inaccessible within the secondary structure of the mRNA, thus preventing translation of the structural gene. The addition of tetracycline causes a conformational change in the mRNA, perhaps as a result of slowing translation of the leader sequence, which uncovers the RBS2 site. Translation of the structural region follows, leading to the expression of the tetracycline efflux pump.

9.4.2 Quinolones

Although the major forms of resistance to the quinolones depend on the mutations in DNA gyrase and topoisomerase IV previously described, an energy-dependent efflux of quinolones is increasingly found in Gram-positive and Gram-negative bacteria. A pump-mediated efflux of fluoroquinolones has also been reported in the urogenital pathogen *Mycoplasma hominis*. In *Staphylococcus aureus*, the *norA* gene encodes a multidrug efflux pump protein NorA belonging to the MFS family, which extrudes quinolones from resistant cells. Increased expression of mutant forms of *norA*, which may be associated with an increased half-life of the *norA* mRNA, is responsible for the higher levels of resistance compared with wild-type bacteria. It is interesting that inhibitors of the function of the NorA efflux protein, such as reserpine and omeprazole (an inhibitor of acid secretion in the stomach), dramatically improve the activities of ciprofloxacin and norfloxacin against strains of *Staphylococcus aureus* which overproduce NorA. More hydrophobic quinolones, such as trovafloxacin and moxifloxacin, are relatively poor substrates for NorA, which may enhance their antibacterial activity against pump-mediated resistant organisms. Homologous MFS-type quinolone effluxing pumps also occur in *Streptococcus pneumoniae*, *Enterococcus faecium* and *Enterococcus faecalis*.

9.4.3 Azole antifungal drugs

Azole antifungal drugs have become increasingly important because the incidence of serious fungal infections has risen sharply, especially among immunocompromised individuals. Oropharyngeal candidiasis caused mainly by *Candida albicans* is a particularly common and distressing infection in AIDS patients. Although azole antifungal agents, such as fluconazole (Chapter 3), are generally effective against *Candida* infections, their ever-increasing use has led to the emergence of azole-resistant strains of this pathogen. There are several modes of resistance, including a reduction in the affinity of the cytochrome P_{450} component of the 14-α-demethylase target and a substantial increase in the cellular content of this enzyme. However, there is little doubt that efflux pump activity makes a substantial contribution to the problem of resistance to azoles. Clinical isolates of resistant strains of *Candida* that fail to accumulate radiolabelled fluconazole may overexpress several genes that encode multidrug efflux systems, including the *cdr1* and *cdr2* genes (signifying ***Candida* drug r**esistance) which encode members of the ABC transporter family, and *camdr1*, which belongs to the PMF-driven MFS superfamily. Both *cdr1* and *cdr2* cloned from *Candida* into *Saccharomyces cerevisiae* confer resistance to fluconazole, ketoconazole and itraconazole. The *camdr1* gene confers resistance to benomyl and fluconazole. The level of expression of these genes in *Candida albicans*, as indicated by mRNA measurements, may increase dramatically during prolonged fluconazole treatment of persistent infections in immunocompromised patients The observed dual overexpression of *cdr1* and cdr2 suggests that these genes share a common transcriptional regulator. However, the mechanisms underlying the overexpression of efflux pump genes in azole-resistant *Candida albicans* could also include gene amplification, mutations in the promoter regions, *trans*-acting factors, or simply a greater stability of the mRNAs. It should be emphasized that the phenomenon of acquired cellular resistance to azoles almost certainly results from a combination of increased efflux pump activity with the other mechanisms mentioned earlier.

9.4.4 Chloroquine

The prevention and treatment of malaria is now in jeopardy in many parts of the world because of the resistance of malarial parasites to chloroquine, a mainstay against malaria for more than 50 years. It has been clear for some time that the resistance of *Plasmodium falciparum* to chloroquine is associated with reduced accumulation of the drug within the parasite. The discovery of two genes (*pfmdr-1* and *pfmdr-2*) for ABC-type transport proteins in chloroquine-resistant strains of *Plasmodium falciparum* suggested that the reduced accumulation of drug might be due to the activity of broad-specificity efflux pumps in removing chloroquine from the protozoal cytoplasm. However, several lines of evidence cast doubt on this appealing model:

1. Drug resistance does not cosegregate with the *pfmdr* genes in a genetic cross between chloroquine-sensitive and chloroquine-resistant strains.
2. It had been expected that the transport protein would be located in the cytoplasmic membrane in order to fulfil its putative efflux function. In fact, the protein is found mainly in the membrane of the intracytoplasmic food vacuole of *Plasmodium falciparum*. The position of the protein in the membrane suggests that it probably transports substrates, including chloroquine, into the vacuole rather than out of it.
3. Some laboratory strains of chloroquine-resistant *Plasmodium falciparum* have **decreased** levels of expression of *pfmdr-1* rather than the increased levels that might have been expected. However, if *pfmdr-1* encodes a transport protein that promotes the uptake of chloroquine into the digestive vacuole, it can be argued that decreased expression of *pfmdr-1* would give rise to resistance.
4. Increased expression of *pfmdr-1* in clinical isolates from Thailand, a region of highly resistant malaria, is associated with resistance to two other antimalarial drugs, mefloquine and halofantrine, but increased sensitivity to chloroquine.

It seems unlikely, therefore, that increased expression of *pfmdr*-encoded proteins is involved in malarial resistance to chloroquine. More recent investigations have defined another gene, *cg2*, that is closely linked to the chloroquine-resistant phenotype in *Plasmodium falciparum*. A genetic cross between a drug-sensitive strain from Honduras and a resistant strain from Southeast Asia showed that the *cg2* gene derived from the resistant parasite has a complex set of polymorphisms, i.e. mutations, that are closely associated with the resistant phenotype. These specific polymorphisms were also detected in chloroquine-resistant parasites from various locations in Southeast Asia and from Africa, thus strengthening the proposal that mutations in *cg2* are responsible for resistance to the drug. A different set of *cg2* mutations was identified in chloroquine-resistant *Plasmodium falciparum* isolates from South America.

The precise location of the CG2 protein encoded by *cg2* within the malarial parasite has been a matter of some debate, but it is now believed to be close to, but not actually within, the membrane of the digestive vacuole. Although the CG2 protein has no amino acid homology with known drug efflux or ion transport proteins, mutant forms of the protein may nevertheless have some role in limiting the intravacuolar concentration of chloroquine. The discovery of yet another gene, *Pfcrt*, apparently tightly linked to the chloroquine-resistant phenotype, adds further complexity to the challenge of unravelling the mechanism of chloroquine resistance. The function of *Pfcrt* is not known at present, but the deduced amino acid sequence of the encoded protein has many of the features of a membrane protein, suggestive perhaps of a transmembrane channel or pump. It is possible that mutations in *Pfcrt* associated with chloroquine resistance may also in some way limit the access of the drug to its site of action in the digestive vacuole (Chapter 6), either by restricting influx to the vacuole or by drug extrusion. Clearly there is some way to go before a complete explanation of chloroquine resistance is available, and indeed of resistance to other aminoquinoline antimalarial drugs. It seems possible that the products of several genes, including *pfmdr-1*, *pfmdr-2*, *cg-2* and *Pfcrt* may all contribute to the resistant phenotype.

9.5 Other mechanisms of resistance

9.5.1 Ribosomal protection against tetracyclines

Ribosomal protection against macrolides and certain aminoglycosides by enzymically catalyzed covalent modification of ribosomal RNA has already been described. The mechanism of resistance to tetracyclines afforded by ribosomal protection proteins is quite different. Nine ribosomal protection proteins encoded by a series of tetracycline resistance genes, including *tet(M)*, *tet(O)* and *tet(Q)* among others, are widely distributed among Gram-positive and Gram-negative bacteria, although they appear to be absent from some Gram-negative enteric species. In contrast to efflux-mediated resistance, the degree of resistance to tetracyclines conferred by ribosomal protection proteins is relatively modest although broader in scope, and includes minocycline. The ribosomal protection proteins have significant amino acid sequence homology with the elongation factors EF-Tu and EF-G which are involved in protein biosynthesis (Chapter 5). The highest degree of homology is found with the GTP-binding domain of the elongation factors. One of the few ribosomal protection proteins so far studied in detail, Tet(M), allows aminoacyl tRNA to bind to the ribosomal A site in the presence of concentrations of tetracycline that would normally inhibit this process. Tet(M) apppears to displace the antibiotic from the ribosome. The favored explanation for the action of the ribosomal protection proteins is that they bind to the ribosome and induce a conformational change which disrupts the interaction between the tetracycline molecule and its binding pocket formed by the head of the 30S subunit at the A site (Chapter 5). GTP hydrolysis catalyzed by the ribosomal protection proteins may energize the conformational change. X-Ray crystallographic data will be needed to substantiate this model of ribosomal protection. There is some evidence that ribosomal protection is inducible by tetracyclines.

9.5.2 Vancomycin

Vancomycin (Chapter 2) has been described as the 'last chance' antibiotic because it is the only drug effective against methicillin-resistant *Staphylococcus aureus* and β-lactam-resistant enterococci. Unfortunately, vancomycin-resistant enterococci are appearing in hospitals in many parts of the world. The *vanA* vancomycin resistance gene cluster in *Enterococcus faecalis* transfers to *Staphylococcus aureus* by conjugation under laboratory conditions and expresses high-level resistance, although as we shall see, *vanA*-mediated resistance is probably not (yet) a major factor in staphylococcal resistance to vancomycin. Details of the mechanisms of vancomycin resistance are of considerable importance in planning appropriate countermeasures.

The antibacterial activity of vancomycin hinges on its ability to bind to the D-alanyl-D-alanine terminal of the peptidoglycan precursor of the cell wall, thereby blocking the activity of the transglycolase essential for the synthesis of the peptidoglycan sacculus (Chapter 2). Enterococcal resistance to vancomycin rests on an unusual strategy in which the terminal D-alanine of the peptidoglycan precursor is replaced by an α-hydroxy acid, D-lactate. The affinity of vancomycin for the D-alanyl-D-lactate terminal is 1000-fold less than for the D-alanyl-D-alanine terminal of susceptible bacteria. A gene cluster harboured by transposon Tn1546 confers vancomycin resistance in enterococci and encodes five proteins. VanH is a dehydrogenase that converts pyruvic acid to D-lactic acid. A ligase, VanA, catalyzes the formation of an ester bond between the D-alanyl residue and D-lactate; and a third enzyme, VanX, is a DD-peptidase that hydrolyzes D-alanyl-D-alanine, thereby virtually eliminating the synthesis of peptidoglycan precursors with D-alanine terminals. The DD-peptidase activity of a fourth protein, VanY, removes the terminal D-alanine residue of any residual normal peptidoglycan precursors that are produced despite the attention of VanX. VanY enhances but is not essential for vancomycin resistance. The fifth protein, VanZ, confers resistance to the related antibiotic teicoplanin by an unknown mechanism.

The drastically reduced affinity of vancomycin for the terminal D-alanyl-D-lactate compared with that for D-alanyl-D-alanine stems from the elimination of a critical hydrogen bond. The amidic NH group of the D-alanyl-D-alanine linkage contributes one of five hydrogen bonds involved in the binding of vancomycin to the dipeptide. This bond is lost when the amide bond is replaced by the oxygen-containing ester link in D-alanyl-D-lactate.

The sophisticated machinery for vancomycin resistance is regulated at the transcriptional level by a two-component system. VanS is believed to be a sensor protein associated with the cytoplasmic membrane that both detects the presence of vancomycin and controls the phosphorylation of an activator protein, VanR, required for the transcription of an operon containing the *vanH, vanA* and *vanX* genes. Phosphorylation of VanR reduces its affinity for the promoter DNA of the *vanHAX* operon. A current model of the inducibility of vanocmycin resistance by vancomycin and other glycopeptide antibiotics suggests that the VanS sensor controls the phosphorylation level of VanR. The presence of glycopeptide antibiotics in the bacterial environment leads to increased phosphorylation of VanR, which in turn permits a higher rate of transcription of the *vanHAX* operon. Just how VanS detects the presence of antibiotic is not known, but possibly an accumulation of peptidoglycan precursors caused by the antibiotic may be involved rather than a direct interaction of the antibiotic with VanS. It is thought that VanS is either a protein phosphatase or protein kinase. The model proposes that in the presence of inducing glycopeptides the activity of VanS is either inhibited or stimulated, depending on whether it turns out to be a phosphatase or kinase. In either event, the phosphorylation level of VanR would be increased.

A semisynthetic derivative of vancomycin, oritavancin (Chapter 2), has significant activity against vancomycin-resistant enterococci. The presence of the *p*-chlorobenzyl side chain in oritavancin, together with the highly dimerized structure of this antibiotic, is thought to enhance anchoring of the drug to the cytoplasmic membrane, thereby increasing the effective drug concentration at the target site and partially offsetting the decreased affinity for the D-alanyl-D-lactate terminals.

Antibiotic 'trapping' in vancomycin-resistant *Staphylococcus aureus*

Contrary to earlier expectations, the mechanism of vancomycin resistance in most clinical strains of *Staphylococcus aureus* is not due to the *VanA/VanB* phenotype but appears to be associated with a remarkable thickening of the peptidoglycan wall which, it is suggested, effectively traps the antibiotic before it reaches the cytoplasmic membrane. The meshwork of the outer layers of thickened peptidoglycan is said to be 'clogged' by the trapped vancomycin molecules themselves, thus further hindering the inward diffusion of antibiotic from the external medium. The cell walls of many VRSA strains from several countries around the world were found to have a mean thickness of 31.3 $\pm$ 2.6 nm compared with 23.4 $\pm$ 1.9 nm in vancomycin-susceptible strains. Despite this striking association between vancomycin resistance and the thickening of the wall of *Staphylococcus aureus*, an actual reduction in the access of the antibiotic to its target site has yet to be convincingly demonstrated.

How does the increased thickness of peptidoglycan arise? Current evidence points to a reduction in the turnover of peptidoglycan associated with reduced activity of the autolysin enzymes that normally remove the older, outer layers of peptidoglycan as the bacterial cell grows. This resembles the long-recognized phenomenon of tolerance in the pneumococci to inhibitors of cell wall biosynthesis. In antibiotic-susceptible pneumococci, the drugs indirectly deregulate the normal control of autolysin activation. Increased autolysin activity synergizes with inhibition of cell wall biosynthesis by causing more rapid dissolution of the peptidoglycan. Mutations in the regulatory control of autolysin in antibiotic-tolerant pneumococci block the signaling mechanism which triggers increased autolysin activity in antibiotic-challenged susceptible bacteria. A range of such mutations has been identified in clinical strains of vancomycin-tolerant pneumococci. The genetic basis of vancomycin resistance in *Staphylococcus aureus* is currently under intensive study.

9.6 Drug resistance and the future of antimicrobial chemotherapy

The development of effective drugs against microbial infections is undoubtedly one of the outstanding successes of twentieth-century science and medicine. However, as we have seen, the whole therapeutic enterprise is threatened by the relentless rise of drug-resistant bacteria, fungi, viruses and protozoa. Mechanisms for eliminating naturally occurring antibiotics by enzymic inactivation and efflux pump systems existed long before the chemotherapeutic era. These, in combination with the mutability of micro-organisms, their high replication rates, and especially in bacteria, their ability to exchange and acquire new genetic material, pose enormous practical and intellectual challenges to scientists and physicians. Some authorities have raised the grim prospect of a time not too distant when the chemotherapy of infectious disease may fail completely in the face of drug-resistant organisms. However, while treatment failures already occur in specific situations, the strategy detailed here may postpone or even avert a global collapse of antimicrobial chemotherapy.

1. The probability of infection itself can be reduced by (a) insistence on high standards of hygiene in hospitals and nursing homes; in educational institutions and places of employment, entertainment and hospitality; and in the manufacture, preparation and cooking of foods, and (b) the further development and vigorous use of prophylactic vaccines to prevent infection.
2. Undoubtedly the trivial, sometimes irresponsible use of antimicrobial drugs in human and veterinary medicine and in agriculture has fostered the emergence and spread of resistant organisms. Even the medically respectable use of antimalarial drugs as prophylactic agents to protect people in malarious zones has contributed to the emergence of drug-resistant malarial parasites. Publicity and a growing awareness of drug resistance has encouraged progress towards a more restrained use of antimicrobial drugs, but their ready availability without medical supervision in many parts of the world remains a serious cause for concern.
3. The therapy of infections should be designed to ensure maximum effectiveness in terms of the choice of drug or combinations of drugs, the dose levels, dosing frequency and duration and finally, patient compliance with the agreed treatment regime. Whenever possible, the objective must be to eliminate the infecting organisms by a combination of direct chemotherapeutic attack and the activity of the patient's immune defences. The need to minimize the numbers of infecting organisms by chemotherapy is of overwhelming importance in immunocompromised patients. Sensitive techniques for monitoring the viral load in AIDS patients have a major role in managing the use of combinations of anti-HIV drugs. Unfortunately, these techniques are often not available in many developing areas of the world where AIDS is now endemic.
4. Despite the measures listed above, there will certainly be a continuing need for new drugs to combat drug-resistant infections. Where the biochemical mechanisms of resistance are understood, ingenious drug design and skillful chemical synthesis may deliver further successes comparable with the development of novel β-lactams. The elucidation of the biochemical systems essential for microbial survival provides new targets for chemotherapeutic attack and will continue to occupy many scientists in academia and the pharmaceutical and biotechnology industries. The ingenious application of molecular genetics to micro-organisms capable of antibiotic synthesis holds the promise of creating novel structures with improved antimicrobial activities. Knowledge of the molecular genetic basis of drug resistance may also be turned to good effect in devising agents to hinder the emergence and spread of resistant micro-organisms.
5. The application of agents to boost the immune defences of patients during infections has so far been relatively limited and restricted largely to immunocompromised individuals. Such treatments are generally expensive compared with antimicrobial drugs, but the ability of biotechnology to produce the many proteins involved in immunocompetence may eventually be valuable in the management of infectious disease.

Further reading

Balkis, M. M. *et al.* (2002). Mechanisms of fungal resistance. *Drugs* **62**, 1025.

Bush, K., Jacoby, G. A. and Medeiros, A. A. (1995). A functional classification scheme for β-lactamases and its correlation with molecular structure. *Antimicrob. Agents Chemother.* **39**, 1211.

Chopra, I. and Roberts, M. (2001). Tetracycline antibiotics: mode of action, applications, molecular biology and epidemiology of bacterial resistance. *Microbiol. Molec. Biol. Rev.* **65**, 232.

Clave, F. and Hance, A. J. (2004). HIV drug resistance. *N. Engl. J. Med.* **350**, 1023.

Golemi-Kotra, D. *et al.* (2003). Resistance to β-lactam antibiotics and its mediation by the sensor domain of the transmembrane BlaR signaling pathway in *Staphylococcus aureus*. *J. Biol. Chem.* **278**, 18419.

Helfand, M. S. and Bonomo, R. A. (2003). Beta-lactamases: a survey of protein diversity. *Curr. Drug. Targets* **3**, 1568.

Hiramatsu, K. (2001). Vancomycin-resistant *Staphylococcus aureus*: a new model of antibiotic resistance. *Lancet, Infect. Dis.* **1**, 147.

Huovinen, P. *et al.* (1995). Trimethoprim and sulfonamide resistance. *Antimicrob. Agents Chemother.* **39**, 279.

Larson , L. L. and Ramphal, R. (2002). Extended spectrum β-lactamases. *Semin. Respir. Infect.* **17**, 189.

Majiduddin, F. K., Materon, I. C. and Palzkill, T. G. (2002). Molecular analysis of beta-lactamase structure and function. *Int. J. Med. Microbiol.* **292**, 127.

McKimm-Breschkin, J. *et al.* (2003) Neuraminidase sequence analysis and susceptibilities of influenza virus clinical isolates to zanamivir and oseltamivir. *Antimicrob. Agents Chemother.* **47**, 2264.

Mitchell L. S. and Tuomanen, E. L. (2002). Molecular analysis of antibiotic tolerance in penumococci. *Int. J. Med. Microbiol.* **292**, 75.

Murray, I. A. and Shaw, W. V. (1997). *O*-Acetyl transferases for chloramphenicol and other natural products. *Antimicrob. Agents Chemother.* **41**, 1.

Nordmann, P. and Poirel, L. (2002). Emerging carbapenemases in Gram-negative aerobes. *CMI*, **8**, 321.

Normark, B. H. and Normark, S. (2002). Evolution and spread of antibiotic resistance. *J. Int. Med.* **252**, 91.

Schmitz, F. J. *et al.* (2002). Activity of quinolones against Gram-positive cocci: mechanisms of drug action and bacterial resistance. *Eur. J. Clin. Microbiol. Infect. Dis.* **21**, 647.

Smith, C. A. and Baker, E. N. (2002). Aminoglycoside antibiotics resistance by enzymatic deactivation. *Curr. Drug Targets Infect. Dis.* **2**, 143.

Sugantino, M. and Roderick, S. L. (2002). Crystal structure of Vat(D): an acetyltransferase that inactivates streptogramin Group A antibiotics. *Biochemistry* **41**, 2209.

Ursos, L. M. B. and Roepe, P. D. (2002). Chloroquine resistance in the malarial parasite *Plasmodium falciparum*. *Med. Res. Rev.* **22**, 465.

Van der Wouden, E. J. *et al.* (2001). Mechanism and clinical significance of metronidazole resistance in *Helicobacter pylori*. *Scand. J. Gastroenterol.* Suppl. **234**, 10.

Yu, S. *et al.* (2003). Reduced affinity for isoniazid in the S315T mutant of *Mycobacterium tuberculosis* KatG is a key factor in antibiotic resistance. *J. Biol. Chem.* **278**, 14769.

Zgurskaya, H. I. (2002). Molecular analysis of efflux pump-based antibiotic resistance. *Int. J. Med. Microbiol.* **292**, 95.

Index

Index